THE WORLD'S BEACHES

THE PUBLISHER GRATEFULLY ACKNOWLEDGES THE GENEROUS CONTRIBUTION TO THIS BOOK PROVIDED BY THE SANTA AGUILA FOUNDATION.

THE WORLD'S BEACHES

Orrin H. Pilkey
William J. Neal
Joseph T. Kelley
J. Andrew G. Cooper

UNIVERSITY OF CALIFORNIA PRESS
Berkeley   Los Angeles   London

CONTENTS

FOREWORD BY THE DONOR, THE SANTA AGUILA FOUNDATION

A beach is not just a pile of sand for us to enjoy, and we hope this book will unveil some of the mysteries of this narrow strip of nature caught between land and sea. Beaches are the most dynamic features on earth, constantly changing shape and providing vital ecological functions and a home to environments of amazing biodiversity. Understanding the importance of the beach's role vis-à-vis the land, the nearshore, and the ocean and its biodiversity is crucial to its protection and preservation.

Sadly, the beauty of our coastlines and the survival of their ecosystems are under threat. America is now facing the repercussions of the BP oil spill, the largest man-made natural disaster in U.S. history. But spills are a global issue, as recently demonstrated in China, Brazil, Nigeria, Mexico, and France. Pollution, overdevelopment, poor coastal management, and constant human interference, including beach-sand mining and seawall construction, endanger coastlines and beaches everywhere. The rise of the sea level has accelerated natural erosion and will result in a substantial loss of infrastructure. This will have an impact on all living beings, but it is our children and future generations who will be most affected.

The Santa Aguila Foundation is a U.S. nonprofit organization dedicated to the preservation of coastlines around the world. It was created after the foundation's founders witnessed the destruction of the beautiful beaches of Morocco to sand mining, a common practice around the world that is largely ignored by the media and unknown to the public at large. Since then, the foundation has focused its energy on global coastal issues (a reflection of the interconnectedness of the planet) and education; the combination of these two elements makes our work unique. We believe that education is key: the more knowledgeable we become about our beaches and coastlines, the deeper our bond with them and the more firm our willingness to protect these endangered natural habitats.

The mission of the foundation is to raise awareness and mobilize individuals to address the practices that are harming the world's beaches and coastlines; to educate children about the scientific aspects of natural beaches and coastlines and thus empower them to act to protect their coastal environment; and to advocate sensible, science-based policies and regulations that will protect and preserve beaches and coastlines around the world.

The Santa Aguila Foundation is honored to have made *The World's Beaches* possible, via a grant, and to be associated with the authors of this book, all highly respected coastal geologists.

We hope you will enjoy this book about beaches and take as much pride as we do in defending this unique feature of our planet.

This project was made possible thanks to the generous donation of Alvise and Andrea Chiari-Gaggia in memory of their mother, Marcella Gaggia.

Please visit www.coastalcare.org for further information.

The Santa Aguila Foundation

PREFACE

We, the authors of this book, often think we are the luckiest people in the world. We have walked on and looked at beaches all over the world, on all seven continents. With our feet and eyes we study one of the world's most dynamic natural environments. Best of all, the work is part of our job: We study the present as geologists in order to understand the past, and as educators to pass on our global experience to students.

At times we have walked around chunks of ice that were pushed ashore by cold Arctic winds so that they bulldozed beach sand on their way. At other times we have tramped along steaming-hot beaches in the tropics next to rain forests alive with strange noises and filled with beautiful butterflies. Some beaches were remote, tens of miles from the nearest person, while others were lined by subsistence villages full of people who mistook us for "officials" because they could fathom no other reason why we would be there. We have walked along some of the world's great tourist beaches, crowded with sun worshippers escaping from their busy lives in well-to-do societies. Often we have appeared to be out of place, wearing long pants, long sleeves, hats, and boots among the more scantily clad beach goers; and instead of lolling on the beach or enjoying the surf, we often were wandering into the dunes or clambering over seawalls carrying our cameras and notebooks. In striking up acquaintances with the locals and tourists, we have learned much about these beaches that we might not have observed and have discovered much about people's conceptions and misconceptions regarding beaches.

We have seen much and found many things that seem strange; these represent natural riddles to be solved, and some of the questions within the riddles remain.

However, our long-term experience has given us a global perspective in regard to beaches, how they form, how they evolve, and how they are similar but different. To us it seems fortuitous, but our coming to the beaches professionally has corresponded more or less with a global rush to the shore, at least in the Western world. Suddenly our work became a bit less academic and a bit more practical and important to society. Expensive houses began to fall into the sea, and seawalls began to sprout like weeds in a garden. One of the first lessons that became apparent was that the price to be paid to protect buildings with seawalls was the eventual loss of the beach. All over the world, planners and politicians wanted to know just how beaches worked and what they could do to save them and the houses next to them.

Meanwhile, in the midst of this societal maelstrom over the "erosion problem," we learned much about the little things that make beaches what they are. We came to know why some beach sand is soft, why some beaches sing when you walk on them, and why some beaches have dark rings on their surface and tiny holes scattered far and wide. We have lived and worked with the scientists who figured out how old (or young) beaches are and how they began and evolved.

All of us are professors, and we have brought students on field trips to beaches for many years. We have found that people are fascinated when they are asked to view a beach as something other than a strip of sand to play on. The features of beaches, large and small, and the mechanics of beaches, explained over the noise of the surf while standing on a breezy shoreline, kept the attention of even the most desultory or distracted student. The leap from those experiences to this book was a short one.

We four authors are friends of long standing who enjoy working together. Starting in 1976, William Neal and Orrin Pilkey began writing and editing coastal books over a span of twenty-seven years, resulting in the twenty-two-volume Living with the Shore series, published by Duke University Press. These state-specific books focus on the hazards of beachfront living. William Neal, an emeritus professor of geology, was a longtime faculty member of Grand Valley State University in Allendale, Michigan, just a dozen or so miles from the shore of Lake Michigan in Great Lakes country. Orrin Pilkey is a retired professor from Duke University, in North Carolina, where he founded the Program for the Study of Developed Shorelines. He coauthored two early books on shoreline problems (*The Beaches Are Moving* and *How to Live with an Island*) that set the groundwork for the Living with the Shore coastal hazard series, and more recently he authored books on the U.S. Army Corps of Engineers and the American shore, global barrier islands, the perils of mathematical modeling, and the sea-level rise. Joe Kelley authored two of the books in the coastal hazard series (Maine and Louisiana). He wrote the Louisiana book in the early part of his career, while he was a faculty member at the University of New Orleans. After moving back to his home state of Maine, Kelley was the state of Maine's coastal geologist for years before he became a professor at the University of Maine. Currently he is the chairman of the Department of Earth Sciences. Andrew Cooper spent a decade in South Africa, living in Durban and observing the coasts of much of southern and East Africa. He eventually moved back to his home in Northern Ireland, joining the faculty at the University of Ulster in Coleraine, where he now heads the Coastal Studies Research Program. Cooper has published numerous coastal studies and has done joint research on a number of coastal issues with both Pilkey and Kelley.

Among us we have well over one hundred years of coastal geology experience. That statistic suggests that we also have spent a good deal of time away from home in our work, although we often brought our families along. Perhaps that explains why our collective marriages have endured for a total of more than 150 years, and we are now counting our grandchildren, and even great-grandchildren. Sometimes the questions posed by our children (thirteen among us) brought our focus to particular beach features. We write so our children and their children can understand and enjoy the beaches as much as we have. The Earth's future is theirs.

The first part of this book, five of the thirteen chapters, begins with a brief look at the role that beaches have played in history over thousands of years. We then turn to the science of beaches, how their study has developed, particularly during the twentieth century, leading to various classifications of coasts, shorelines, and beaches. Then we turn to the materials of beaches, and how the waves and sand interact, what happens in storms, and considerable discussion about why beaches are all different. Part II of the book, chapters 6 through 11, provides guidance in how to read a beach, how to explain what we can see on a beach, and what beach surfaces tell us about how beaches work.

Chapters 12 and 13 in Part III explore the threats that beaches face today: coastal overdevelopment, pollution, oil spills, the impacts of coastal engineering, and especially the rising sea level. For example, while we were in the final manuscript preparation in the early months of 2010, a major storm hit the west coast of France, ultimately resulting in fifteen hundred houses being condemned (and a major relocation). After that, the oil-well drilling-platform disaster in the Gulf of Mexico threatened the U.S. Gulf Coast with an oil spill that became the largest ever for the United States. The small, almost unnoticed, reports of beach mining, refuse accumulation on tourist beaches, development controversies, and the stories of the sea-level rise on beaches here and there continued to come in, daily reminders of the varied threats to beaches. There is no question that the beaches that our grandchildren will play on will be different from ours. The important question is whether they will be better or worse.

We have a large number of people to thank for helping us with the book, more than we can list here. Of course, in summarizing the nature of beaches we stand on the shoulders of a dozen prominent international scientists who preceded us. These pioneering individuals came from all over the world, including Australia, Germany, New Zealand, Russia, the United Kingdom, and the United States. Miles Hayes, longtime global beach watcher and the "king" of barrier island science, contributed some outstanding photos to our book (and discussed them at length with us). Charles Pilkey, artist son of Orrin, created the various line drawings and illustrations. Norma Longo provided essential assistance as an overall organizer, editor, file clerk, adviser, and researcher for this book. Numerous individuals provided us with photos. We extend special thanks to Angela Hessler, Joe Holmes, and Mark Luttenton for assistance in photography, and we note that the photos from northern Alaska beaches were contributed by Owen Mason, Puget Sound photos by Hugh Shipman, and Antarctica photos by Norma Longo. Siberian photos were the result of a field trip arranged by Wally Kaufman, who provided important input when this book was in its formative stage.

The manuscript was improved as a result of Duncan Fitzgerald's careful review, for which we thank him, but any errors that might remain are those of the authors alone. We especially thank our editors, Jenny Wapner, Lynn Meinhardt, and Hannah Love, along with the University of California Press for seeing us through the production of this book. Encouragement is a driving force in any work, and there are many people and programs dedicated to protecting beaches. For this effort, we extend our gratitude to Eva and Olaf, to the Program for the Study of Developed Shorelines, Western Carolina University, and to the Santa Aguila Foundation. The production of this fully color-illustrated book was made possible by the Santa Aguila Foundation, and we encourage readers who are concerned about the conservation of beaches to visit both the Coastal Care and the Program for the Study of Developed Shorelines Web sites.

PART I

THE GLOBAL CHARACTER
OF BEACHES

Beaches are at the top of the list of Earth's natural attractions, drawing millions of visitors in all parts of the world. These wedges of sand and gravel, held against the shoreline by ocean energy, are among the most dynamic of natural environments. Beaches also are something of a natural riddle; though they all share commonalities, each one is different. The goal of this book is to solve that riddle by examining the dynamic processes that produce beaches, the character of the materials that make up beaches, and the great variety of physical and biological features that are found on beaches.

The shoreline boundary between land and sea is one of nature's longest and most fascinating features, and a significant portion of this great feature consists of beaches. Eric Bird, in his 2008 book *Coastal Geomorphology*, states that the global shoreline is on the order of more than 620,000 mi (1 million km) in length. Sandy beaches probably account for just over one-third of this great length. The following five chapters view beaches on a global scale, beginning with a short historic overview, then proceeding through global classifications of beaches, the sources of beach sediments, the shaping of these materials by waves, tides, and currents, and an outline of the various parts of beaches and how they differ regionally.

When you finish, you'll never again look at the beach in the same way.

1

A WORLD OF BEACHES

Beaches are a treasure—cherished by most, exploited by some, enjoyed by all. Beaches are places for recreation, contemplation, renewal and rejuvenation, communing with nature, and sometimes, while staring out to sea, thinking about our place in the universe. On beaches we swim, surf, fish, jog, stroll, or just lose ourselves in the wonder of where the land meets the sea. Yet for all of our interaction with beaches, few of us understand them: why they are there, how they work, why they show so much variety in form and composition, and why they can undergo dramatic changes in a matter of hours.

CROSSROADS OF HISTORY

Humans have been crossing beaches since the dawn of time, and beaches have been critical to human history and development, as they still are. Unfortunately, much of the history of beaches has to do with invasions, but discovery was also part of the human tide that traversed beaches through history. Julius Caesar landed on Deal Beach near Dover when he invaded Britain in 55 B.C., fifteen hundred or so years before Columbus landed in the New World. In A.D. 1001, Leif Ericksson was the first European to set foot on a beach in Vinland (Newfoundland). King Canute sat on his throne on a beach in 1020 and ordered the tides to come no closer, an early object lesson to demonstrate to his subjects that no man, not even the king, has authority over the sea. The Normans crossed the beach at Hastings, England, in 1066 to defeat the English. The Mongols

A beautiful cliffed shoreline of volcanic rocks in the Aleutian Islands, Alaska. Two small pocket beaches are visible at the base of the cliff. The material of these beaches ranges from sand size to boulders.

crossed the beach at today's Fukuoka, Japan, in 1281 to be defeated by the divine wind, a typhoon that destroyed the invasion fleet. The Spanish Armada of 1588 met a similar fate in their attempt to invade England when a great storm blew the surviving ships onto the rocky coasts of the British Isles. Many of the survivors and much debris and treasure washed up on Ireland's beaches. Columbus planted the Spanish flag and a cross in 1492 on the beach at San Salvador in the New World, to the amazement of the natives. In 1519, Hernán Cortés and six hundred of his men crossed the beaches of the Yucatán Peninsula on his way to conquering the Aztec Empire. Australians first met Aborigines on a beach in 1606. In 1619, a Dutch vessel landed twenty slaves on a beach in Chesapeake Bay, marking the beginning of African slavery in America. In 1620, the Pilgrims disembarked in the New World next to a large rock on the beach now known as Plymouth Rock. In 1659, Robinson Crusoe is said to have crawled across the beach on an uninhabited island off the Orinoco River, in northern South America, where he remained for twenty-eight years. The great explorer Captain Cook met natives on the beach in Hawaii (the Sandwich Islands), where they killed him in 1779. And Darwin met naked Patagonians on a cold beach in Tierra del Fuego in 1833.

In 1915, nearly 330,000 total casualties occurred on or very near the beaches of Gallipoli, Turkey, as the Turks beat back the invading Allied forces. Will Rogers died when his plane crashed on takeoff from a beach near Barrow, Alaska, in 1935. And the beach at Dunkirk, France, in 1940 was the scene of the spectacular rescue of the defeated British Expeditionary Force in World War II. In 1944, the direction of the armies reversed as the Allies invaded Europe across the beaches of Anzio, Italy, and then Normandy, France. In the same time interval, beaches across the Pacific were killing fields as the Allies moved against the Japanese, culminating in the atomic bomb tests at the Bikini Atoll, the namesake for the bikini bathing suit, introduced by a Frenchman in 1946. The largest oil spill in history soiled the beaches of Kuwait and Saudi Arabia in 1991, when Iraq purposely released oil to frustrate beach landings by U.S. Marines in the Gulf War. In 2010, the BP Deepwater Horizon blowout in the Gulf of Mexico became the largest oil spill ever to occur in North America.

Upper A busy summer beach scene in Fréjus, France. The beach is backed by a seawall designed to protect buildings during storms.

Lower People on the beach in Kuwait. More than half of the people on this beach appear to be nonswimmers, but there are many ways to enjoy a beach. Photo courtesy of Miles Hayes.

AVENUES OF COMMERCE

Having been erased by erosion and flooded by the rise in sea level, archaeological sites are less common on today's beaches than they were in the past, but we can guess that early humans used the beach in much the same way as today's third world coastal communities and subsistence cultures do. The beach was their land road, and just as for today's subsistence societies, from the Arctic to the tropics, living next to the beach is living next to one's main source of food. Places near the beach were also dump sites for garbage. Termed "middens" by archaeologists, massive piles of shells are common in many coastal settings near beaches and tidal flats where food resources were common. Today on Bazaruto Island, Mozambique, and in other coastal subsistence societies, local people still contribute to growing shell middens.

From the North Slope of Alaska to the tropical shores of the Pacific in Colombia, beaches continue to be workplaces and storage places for fishing boats, and spaces for

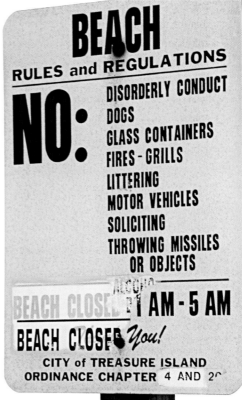

Dubai beach sign noting that only women can swim here on certain days. Women can be accompanied by males, provided they are four years of age or younger. On tourist beaches in Dubai, however, *almost* anything goes!

A sign posted at a beach in Treasure Island, Florida. As traffic increases, more traffic regulations are emplaced. The same is true for the density of beach use. This sign is typical of the increasing need to regulate multiple uses on beaches, but sometimes it looks like having fun is prohibited!

net- and fish-drying racks. In the tropics, sea breezes provide relief from the heat and help reduce malarial mosquitoes. The beach itself is a resource for construction material and for whatever bounty the sea delivers. The people of such communities live by the sea by necessity; it is their means of life. With a vista to see who is approaching, a beach provides security. But living next to the beach, particularly on low-lying coasts, presents great risks, as demonstrated by the great tsunami of 2004 that roared across thousands of miles of Indian Ocean beaches and killed 225,000 people—including those who were there by necessity and those who were there by choice.

In contrast to beaches that support subsistence cultures, urbanized shores are mostly characteristic of first world countries. The combination of the shore as a place of commerce and the shore as a place of leisure is probably as old as humankind. The ruins of Roman and Greek villas by the sea attest to a very early resort mentality, whereas ancient Peruvians built massive temples and dug grave sites near their beaches. It was not until the nineteenth century that beaches became a greater focal point for technological and recreational development. In 1801, the first American advertisement for a beach resort (Cape May, New Jersey) appeared in the Philadelphia *Aurora*. In 1845, the Sanlucar de Barrameda beach horse race began in Spain, and beach horse races in Laytown, Ireland, commenced in 1876. The first successful transatlantic telegraph cable, completed in 1866, crossed the beach at Heart's Content, Newfoundland, in the west, and at Valentia Island, Ireland, in the east. In 1898, gold was mined on the beach at Nome, Alaska. In 1903, the speed of a horseless carriage was timed on the beach at Daytona Beach, Florida. Beginning in 1905, Duke Kahanamoku rejuvenated the Polynesian sport of surfing, which the Hawaiian missionaries had halted earlier for being ungodly. In 1927, the same year that Charles Lindbergh landed the *Spirit of St. Louis* on the beach at Old Orchard Beach, Maine (the airport was fogged in), beach volleyball was introduced to Europe in a French nudist camp. "Beach

The site of Eric the Red's Viking village on the northern tip of Newfoundland. The mounds of earth record the sites of individual buildings that were once close to the shoreline. The sea level is dropping here because the land is rebounding from the removal of the weight of the former ice sheets. The marsh in the background was once a small harbor and is uplifted and now preserved.

Beachfront condominiums in Dunkirk, France. This peaceful scene belies the violence that occurred here in 1940 as the British army avoided annihilation and escaped back to England. The small bunker in the foreground is all that remains to remind us of the historic event that occurred here. The tide range at Dunkirk is nearly 20 ft (6 m), and at low tide the beach is often more than 435 yd (400 m) wide, with four to five sand ridges on the intertidal beach.

Upper An aerial view of a heavily oiled beach along the Saudi Arabian shoreline. The oil was spilled purposely by the Iraqis in January 1991, during the Gulf War, in order to prevent a seaborne invasion by coalition troops. This was by far the largest oil spill in history, amounting to as much as 520 million gallons. Although oil now is no longer visible on the surface of the beach, concentrations of oiled sand can be found within a foot or two of the surface. Photo courtesy of Miles Hayes.

Lower A pool of oil on Grand Terre Island, Louisiana, in May 2010, one of the early results of the BP Deepwater Horizon oil spill in the Gulf of Mexico, the largest oil spill in U.S. history. The horrific oil pool here provides a beautiful reflection of the clouds. Photo courtesy of Adam Griffith.

music" started in 1945. In 1953, Deborah Kerr and Burt Lancaster made love on a beach (Halona Beach, Hawaii) in *From Here to Eternity*. The Beach Boys rock band formed in 1961. In 1963, Frankie Avalon and Annette Funicello starred in the surfing classic *Beach Party*, and in the 1968 movie *Planet of the Apes*, Charlton Heston and Kim Hunter, riding horseback on a beach, discovered the ruins of the Statue of Liberty.

From the post–World War II era to the present, coastal resort communities have experienced rapid growth. This time period also has witnessed the greatest losses to both coastal property and, more significantly, the beaches themselves. The 1962 Ash Wednesday storm along the U.S. East Coast caused beach loss so significant, particularly in New Jersey, that it precipitated the U.S. national beach nourishment program. This approach has been widely adopted, leading to many artificial beaches internationally (see chapter 12).

NATURE'S MOST DYNAMIC ENVIRONMENT

Independent of the fact that beaches have played a significant role in history, these natural systems are quite amazing and unique in their behavior. Beaches are arguably the most flexible and dynamic features in nature. If we did not know better, we might think that beaches are living creatures. They do things that make sense: Beaches protect themselves during storms by hunkering down and flattening, which makes the storm waves dissipate their energy over a broadened surface. If the sea level rises, the beach does not disappear. Instead it moves up and back toward the land, apace with the water-level rise.

Viewed from the air, beaches are the thin line that marks the boundary between terra firma and the great blue expanse of the ocean. This graceful winding line is not fixed; it changes constantly. The line waves back and forth, both landward and seaward, although nowadays the line usually is moving landward by a process called *shoreline retreat* (also called erosion or migration). It is fair to say that most of the world's beaches are retreating, partly in response to a rising sea level.

The beach changes its shape constantly, whether viewed in cross section or in profile. The alert beach visitor who comes to the shore in different seasons may see large differences. Some changes may occur within a few hours during a storm, and some may manifest over the course of months, as the beach responds to seasonal differences in wave energy. When engineering structures are put in place to hold the shoreline still and protect buildings, the beach behaves quite differently than it does in its natural state. Usually it becomes narrower and over time may even disappear altogether.

Beaches range in color from white, as on the coral beaches of Pacific atolls, to pink in Bermuda, to yellow-brown on southeastern U.S. beaches, to black on volcanic islands. A few beaches have strange colors (see chapter 3); for example, Papakolea Beach, Hawaii, where the mineral olivine is concentrated, has green sand, and Northern Labrador has red beaches, which reflect the color of abundant garnet.

Even smaller features of beaches, those just beneath our feet, change very frequently and rapidly with each breaking wave's swash and backwash, with each gust of wind, with whatever organisms are working on or within the sand. These various beach surface features, referred to

Upper Known as the Glidden midden, this accumulation of shells was left by years of Native American shell fishers at Damariscotta, Maine. The midden is eroding and providing an abundance of shells to the beach.

Lower Scallop shells discarded on the beach at Portavogie, Northern Ireland. These shells came from a seafood-processing plant that started operations about twenty-five years ago. The shells are still accumulating in a modern midden.

as *bedforms*, give particular character to the beach and are as fascinating as the shells and the flotsam and jetsam that are often the focus of our beachcombing. These features often raise the most questions in terms of what, how, why, and when, as beach aficionados attempt to "read" the beach.

A healthy beach is a dynamic beach, but humans tend to think of beaches as permanent in their location, and they dislike natural features that move about, particularly

Peruvian beach at Chan Chan, near the city of Trujillo. The mound of sand is believed to be the ruins of a large temple built close to the shoreline, and hundreds of robbed graves (depressions) are visible in both the foreground and background. This Peruvian coast is a desert environment, which means that vegetation plays a much smaller role in forming and maintaining sand dunes than in more temperate climates.

when people have placed buildings in the path of such movement. In fact, the only real enemy that beaches have is us.

From remote Eskimo villages in Siberia or the barrier island villages in Nigeria, to shoreline urban developments such as the Gold Coast of Australia (with its eighty-five-story beachfront condo) or the endless line of high-rises on Saint Petersburg Beach, Florida, the greatest fear of all beach inhabitants is the landward movement of the beach. On the Gold Coast and on Saint Petersburg Beach, Florida, communities erect seawalls and make artificial beaches to replace the native sand. In Nigeria and Siberia, where less money is available, houses are often moved back from the beach. Ironically, the result is that beaches often remain more pristine in poor societies than in affluent ones.

VARIED MATERIALS

All of these generalizations about shape, color, surface features, and changes pertain to sandy beaches. Many beaches in the world are made up of sand, but many also consist of gravel (throwing-size pebbles), cobbles (grapefruit-size stones), and even boulders (rocks too big to lift), depending on where the beach material came from. In high, Northern Hemisphere latitudes, many beaches are made of glacial sediment, carried to the site by the now-retreated glaciers. Some rocks on Danish beaches came from Norway, and some material on Irish beaches hails from Scotland, indicating that various beach materials were transported many miles from their original locations. Sand on many beaches originated as rocks that were located hundreds or thousands of miles away and were weathered and transported by rivers. In contrast, the white sand on the beaches of many tropical islands was transported only a few meters from offshore reefs. Some boulder beaches are derived from disintegrating cliffs at the back of the beach, or just "upstream," where the beach connects to an eroding bluff. Gravel beaches also may be derived from concentrations of seashells, coral fragments from offshore reefs, or adjacent streams and rivers that carry mainly gravel. Arctic beaches are commonly gravel.

Halona Beach, Oahu, Hawaii, where Deborah Kerr and Burt Lancaster made love in a scene from the movie *From Here to Eternity*. This is a pocket beach, a common type of beach on volcanic coasts like those of Hawaii. Photo courtesy of Norma Longo.

Microscopic view of Hawaii green sand, composed mainly of the green mineral olivine, a few pink to cream-colored shell fragments, and black sand-size rock fragments.

Sometimes long stretches of shorelines have no real beaches at all but have mudflats instead. Perhaps the most famous such occurrence is the shoreline north of the mouth of the Amazon River in Brazil. It is along this shoreline that Amazon River sediment is transported by waves and currents, and because the river carries very little sand, the "beaches" are broad mudflats, stabilized by mangrove forests, all the way up to Suriname, more than 400 miles away.

Occasionally a beach is virtually hidden from view by logs or seaweed. In Spencer Gulf, Australia, some beaches are completely covered with a thick layer of sea grass washed up from adjacent shallows. In some beaches in northern Brittany, seaweed, formed as a result of overfertilization of nearby farm fields, is sometimes piled a meter deep on local beaches. The rotting seaweed produces toxic gases that have been lethal to animals—not a good recommendation as a tourist hot spot. Beaches off the Mississippi River mouth are covered with plant detritus (salt-marsh straw) from the extensive salt mashes of the delta. So much grass is deposited there that natural gas formed by

A beautiful white carbonate sand beach in the British Virgin Islands. The white color is typical of beaches made up entirely of calcium carbonate skeletons of marine plants and animals.

the breakdown of the organic matter is emitted from holes in the beach.

In remote areas adjacent to the mouths of major rivers, logs and driftwood that float down the rivers virtually cover the beach. At the mouth of the Magdalena River on Colombia's Caribbean coast, for example, beaches to the west of the river mouth are hidden from view by the log cover. Abundant logs on the beaches of Oregon and Washington states and British Columbia, Canada, shift without warning as waves strike them and have proven to be hazardous to beach strollers.

Beaches also have been the landing sites for oil spills. The 1969 oil spill in Santa Barbara, California, provided the United States with a dramatic wake-up call to the problem. Since then, most spills have come not from oil production areas such as Santa Barbara but from shipwrecks (see chapter 12).

OBSTACLE COURSES

While nature throws up obstacles to our trip along the beach in the form of rocky outcrops, boulders, or mazes of driftwood and fallen trees, the larger accoutrements from human activities also provide obstacles. All sorts of objects cross beaches. In the Gulf of Mexico, hundreds of pipelines carrying oil and gas traverse the beaches on their way to refineries and storage facilities. In the latter half of the nineteenth century, many telephone and telegraph cables, some transoceanic, angled off to the ocean under the beach sand. During the Cold War, the United States and the Soviet Union, laid many miles of cable intended to reveal the presence of "enemy" vessels, both on the surface and submerged. One of the listening stations was at Cape Hatteras, North Carolina, where at least once some mysterious cables appeared on the beach after a storm.

Then there are the artifacts of war. Pillboxes, watchtowers, antisubmarine canon emplacements, antiaircraft batteries, forts, antitank and anti–landing craft structures, and searchlight platforms still abound on the beaches of various combatant countries. Many have fallen into the sea or are stranded on the beach. A few have been made into beach houses; they are perfect for storm-resistant dwellings, but given their low elevations they are unsuitable fortresses in which to ride out coastal storms.

Lighthouses all over the world have crossed the beach on their way to oblivion as shoreline retreat caught up with and then passed their sites. In a famous 1999 move, the 3,500-ton Cape Hatteras Lighthouse in North Carolina was moved back 2,000 ft

An Antarctic boulder beach occupied by local inhabitants (gentoo penguins). Beaches here are covered in ice every winter, so winter storms have little impact on beach evolution. Often the beach material on the scarce Antarctic beaches is ice rafted. Since a small iceberg that might be grounded here could drift in from a wide variety of locations, the beach is often made up of a wide variety of rock types. Photo courtesy of Norma Longo.

Intertidal bars (sand waves) on a mainland Gulf of California beach, made by both wave and tidal current action. The tops of the sand waves seen in this aerial photo are light brown in color because the sand has been dried out by wind. They are spaced about 300 ft (91 m) apart.

A mother grizzly bear and her cub grazing on barnacles at low tide on an Aleutian rocky shore. You would not want to go ashore at this location. The beach material here ranges in size from large boulders to pebbles.

Imagine approaching a beach only to find musk oxen preventing your access to the shoreline. These musk oxen are on the beach at Cape Espenberg along the Chukchi Sea, well above the Arctic Circle in Alaska. Photo courtesy of Owen Mason.

(610 m) to save it from storms and the retreating shoreline. The foundation ruin of its predecessor lighthouse resides on the inner continental shelf, just offshore.

The most frequent obstacles one encounters on beaches, however, are the engineering structures that attempt to hold back the sea. Seawalls, fields of groins, and breakwaters, along with other beach construction such as piers, often in a state of failure or ruin, interrupt beaches' natural topography and ultimately cause the loss of the beaches' natural ambiance (see chapter 12).

This eighty-five-story condo on the Gold Coast beach, Australia, may be the highest beachfront building in the world. The problem with high-rises such as this one, and even smaller ones, is that they reduce the flexibility of a community to respond to sea-level rise. Moving an entire resort community of high-rises to higher ground is impossible, and the community is not likely to be defended when the sea-level-rise crisis comes, because funding will go to protect ports, industries, and urban areas.

TO BE OR NOT TO BE

Natural beaches are in fact threatened by at least three major human actions: engineering, mining, and pollution. So many beaches are just elongated engineering projects that one sometimes gets the impression that engineers run beaches. The entire coast of Belgium is lined by seawalls. Miles and miles of the southern coast of Spain are walled. Sometimes the shoreline is lined by sand-trapping groins, which are short walls built perpendicular to the shoreline. Long walls, called jetties, next to inlets and navigation channels sometimes extend more than a mile into the sea. They prevent sand from clogging the navigation channel, but they, along with groins and seawalls, cut off the sand supply that nourishes downdrift beaches. Finally, piles of rock called breakwaters are mounded in the water off beaches to reduce the impact of storm waves on the beaches and the houses landward of them. They perform well in small storms but fail during big storm events, as do the other structures mentioned here.

Fundamentally, engineering applied to a beach is done with the philosophy that preservation of buildings is more important than preservation of the beach. Beaches or buildings; we can take our choice.

Beaches are mined all over the world. The most important product is sand, many tons of which are used in construction projects. There is hardly a beach in the world that has not been mined at one time or another on scales ranging from wheelbarrow loads, as on Pacific atolls, to multiple front-end-loader and dump-truck loads, as in

Morocco (see chapter 3). Elsewhere, just about any valuable mineral that comes down a river is mined from beaches, including gold in Nome, Alaska; diamonds in South Africa and Namibia; iron in California and Oregon; tin in Indonesia; titanium in Australia; zirconium in Brazil; and uranium in India. The minerals sought are usually a component of the "black sands" that commonly include the minerals zircon, cassiterite, monazite, magnetite, ilmenite, limonite, garnet, staurolite, and rutile.

Logs on a beach in Puget Sound, Washington. In forested areas such as this, logs are an integral part of the beach and its ecosystem. If the logs are "cleaned up" (removed), rapid erosion often ensues. Photo courtesy of Tom Terich.

This pipeline in Bay Marchand, Louisiana (Gulf of Mexico), extends from the shore to the offshore oil rigs that can be seen in the distance, and stretches landward across the shoreline. The beach here consists almost entirely of salt-marsh peat covered by seaweed. Often such pipes are buried in the beach and are not visible to the beach stroller.

COMBER'S DELIGHT OR NATURE'S TRASH COLLECTOR?

The term *beachcomber* first referred to eighteenth-century Europeans on South Pacific islands who combed the beaches for edibles and anything else that came ashore. Many of them were sailors who, tiring of the dangerous and difficult life of sailors of that day, jumped ship when the opportunity arose. *Beachcomber* took on the connotation of a vagrant living on the beach. Today, the term refers to a major recreational activity enjoyed all over the world. Beachcombing ranks as one of the top beach activities, and the treasures one finds can contribute to an understanding of some aspects of the beach.

One never knows what a stroll on a beach will reveal. Everything that floats on the surface of the oceans or that is moved landward by waves in shallow water can end up on the beach. The beach is the landward edge of a gigantic ecosystem, and everything that dies and floats, from whales to microscopic organisms and their shells, may eventually make it to the beach. Hunting seashells is perhaps the most celebrated activity on the world's beaches. On Sanibel Island, Florida, shell seekers come out on the beach in the middle of the night with flashlights, seeking shells at low tide and at a time when they will encounter less competition from other shell seekers. On the world's barrier island beaches, shells that appear to be fresh may actually be fossils that found their way to the beach as the island moved landward over former lagoon shell layers. Some shells on beaches are thousands of years old; a few are millions of years old (see chapter 10). Other treasured beach finds include colorful glass net floats, often from far away, colored glass fragments rounded by wave abrasion (called beach glass or mermaid's tears), crab-trap floats, lobster traps, bones and skulls of various marine organisms, turtle shells, and whale bones and baleen. Not-so-treasured beach finds include tar balls, which are the remnants of oil spills, much plastic debris of all sizes, shapes, and colors, freshly broken glass, and rotting carcasses of various marine organisms.

Items from faraway places are found on beaches, and some of these are very strange indeed.

This 8 in (20 cm) cannon, now fallen on the beach along with other defensive structures in the background on Xefina Island, Mozambique, was part of the coastal defenses that protected Maputo, the capital, during World War I. All over the world, implements of war, including pillboxes, watchtowers, and gun emplacements, now reside on the beaches or are entirely submerged because of shoreline retreat.

Coconuts and mangrove seeds from the tropics are occasional components of mid-latitude beaches. In fact, specialized collectors search for sea beans and drift seeds, inventorying their finds and competing for the largest or farthest-traveled of particular species (for example, a Mary's bean, also known as a crucifix bean, has been found approximately 15,000 mi [more than 24,000 km] from its source). Bales of raw rubber, sized for transport in dugout canoes on the Amazon or Orinoco River, have been found on Puerto Rican beaches in the Caribbean. Logs from exotic tree species from hundreds and even thousands of miles away are mixed with logs from local trees.

In the 1950s, thousands of U.S. Navy mops showed up on a very remote Pacific Coast beach off Baja California, Mexico; they were seen by only a few people. In 2008, long reaches of Suffolk, United Kingdom, beaches were piled deep in lumber from a ship that lost its load in a storm. In contrast to the appearance of the mops in Mexico, this beach incident made most of the world's newspapers and TV news programs. We imagine local folks became true beachcombers in these two instances, salvaging mops and lumber. Certainly salvaging is still important on third world shores. Subsistence fishermen in Colombia collect beach sandals (flip-flops) from the wrack and trim the soles into disc-shaped floats for their nets.

Upper left The 3,500-ton Cape Hatteras Lighthouse, on the Outer Banks of North Carolina, was threatened by a retreating shoreline and was moved back 2,000 ft (610 m) from the shoreline in 1999. Built in 1870, the original location was about the same distance back from the shoreline as it is today. The $12 million move was strongly opposed by local politicians, who feared it would draw global attention to the local shoreline erosion problems (it did). Photo courtesy of the U.S. National Park Service.

Lower left A groin in front of beach hotels in Rhodes, Greece, that has clearly widened the beach in one direction (updrift) by trapping the sand, but has caused a sand deficit and beach loss in the other direction (downdrift).

Chairs on the widened beach indicate heavy use of the beach by tourists, but the narrowed beach is much less useful for public recreation. Photo courtesy of Norma Longo.

Upper right Part of the massive seawall that stretches in one style or another along Belgium's entire shoreline. Judging from the wet-dry line, there is little to no beach here at high tide, a global problem with seawalls. When the beach disappears, tourists are entertained by the view from the promenade on top of the wall. Note the World War II bunkers to the right (green belt).

Lower right Seawall with a promenade in Nusa Dua, Bali, Indonesia. Southeast Asia has become a popular beach tourist destination.

In 1991, sixty thousand Nike shoes were released from spilled containers carried on a storm-tossed ship, and many eventually ended up on beaches on all sides of the Pacific Ocean. Upon arrival on the beaches, the shoes were still usable, once cleaned, and beachcombers organized exchange get-togethers during which one could trade a left for a right shoe or a size 9 for a size 12. In 1995, twenty-nine thousand rubber duckies and other bath toys were released into North Pacific waters near the international date line, again from containers washed over a ship's side. Within ten months, the first rubber duckies appeared on Canadian beaches.

Some beach debris is gruesome. In 2008, two young girls playing on an Arbroath beach on the east coast of Scotland found the head of a woman wrapped in a plastic bag. The origin of the head remains a mystery. Also in 2008, over a period of several months, five running shoes washed up on British Columbia beaches, each with a human foot still inside. Four were left feet and one was a right foot. The identities of the owners of the feet also remains a mystery. In the 1970s, a local physician found a piece of shipwreck timber on a North Carolina Outer Banks beach. The piece of cypress wood had two clumps of rust on it separated by a few inches. Close examination of the rust revealed fragments of a fibula and tibia in each. The explanation: The wood fragment was part of a slave ship that sank with its human cargo shackled to ship timbers, unable to escape. More recently, in January 2010, wreckage from an Ethiopian airliner carrying ninety passengers washed up on a Lebanese beach.

Beware of this one. This Portuguese man-of-war has tentacles that can give a devastating sting. Although not true jellyfish in spite of their appearance, these threatening species are found all over the world in warm waters. If spotted on a beach, they should be looked at but not touched, and beachgoers should be aware that they are likely to be in the water column as well, so swimmers and waders should be wary.

In 2008, Hurricane Gustav drowned thousands of the invasive rodent nutria in the marshes of Louisiana. Storm currents eventually washed seven thousand nutria carcasses up on the beach of Waveland, Mississippi. Old-timers in Waveland tell of a hurricane in the distant past that washed dozens of dead cattle onto the same Waveland beach. The cattle had been grazing on the narrow, low barrier islands, just offshore from the community.

Solid waste disposal on land is a problem, but the accumulation of such waste along the global shoreline is reaching a crisis point. Ships that throw trash overboard, communities that put their wastes, liquid and solid, into the sea, the great infusions of debris and trash that are introduced into coastal waters in every hurricane and typhoon, are impacting all marine environments, but especially beaches. The following examples describe only a small portion of this pervasive problem.

The aftermath of Hurricane Ike, which struck the Texas coast in 2008, left tons of refuse on Texas beaches. A month after the storm, the beaches' appearance was described as that of a "dump," and the trashed beaches extended for a hundred miles from the point of the storm's landfall. At the same time, a zone of floating trash some 80 mi (129 km) long was still located offshore, and onshore transport continued to bring this debris to the beaches for months afterwards—not a happy situation for the coastal communities there that depend on beaches for tourism.

Upper left A debris-laden seawalled beach in Newcastle, Northern Ireland, shortly after a storm.

Center left A debris line left along the Galveston, Texas, shoreline after Hurricane Ike in 2008. Even after the beach is cleaned up, debris from such storms may wash up onto the beach for years. Beachcombers often find valuables from destroyed homes, including jewelry, money, and the like. There are a number of urban legends associated with post-hurricane beach scrounging, including the alleged, but never quite verified, discovery of a can containing $100,000 after Hurricane Ike.

Lower left Shoes on a beach in Sefton, England, along with an abundance of razor-clam shells. Shoes are a common component of beach debris, almost as common

as caps and other headwear, blown off the heads of unwary fishermen. It is a little more difficult to explain why shoes are so common on beaches, though forgetful bathers perhaps stay in the water while the rising tide steals their footwear. Even stranger are the five shoes that showed up on beaches on Vancouver Island, British Columbia, Canada, with human feet still attached, a mystery that has yet to be solved.

Upper right Trash on a beach in Kashima, Japan. Much of this trash probably came from offshore, where it was dumped by passing vessels, large and small. Most beaches in Japan are periodically cleaned of such debris.

Lower right Trash on beaches is all too common, and animals rooting in beach garbage follow. These pigs are rooting in the beach litter at Mezquital on an inlet of the Laguna Madre, Tamaulipas, eastern Mexico.

BEACH STUFF

Following is a list of some of the interesting things (excluding the vast variety of shipborne garbage and the wide variety of dead marine organisms) that we (the authors) have spotted over the years as we have wandered about on beaches.

Bikinis and swimsuits

Bottles containing messages

Bricks

Caps and hats (by the hundreds)

Cars and their sundry parts

Coconuts and large seeds, far from
  their origins

Computers

Elephant tusks (fossil and modern)

Fossil shark teeth and other marine
  vertebrate remains

Fossil shells and bones

Glass and cork net floats from
  distant shores

Houses

Native American arrowheads on
  Chesapeake Bay beaches

Kitchen sinks

Life preservers, rings, and rafts

Fishing lines (miles of them) and
  other fishing paraphernalia

Lobster- and crab-trap buoys (by the
  hundreds)

Lobster-trap buoy from Maine on
  an Irish beach

Logs

Marijuana bales

Mops

Navigation markers

Pottery fragments (modern and
  prehistoric)

Bales of raw rubber

Shipwrecks and shipwreck timbers

Spanish soccer ball on a Moroccan
  beach

Stone-age tools on a Portuguese
  beach

Vacuum cleaners

Whales

Whale vertebrae

World War I cannon on a
  Mozambican beach

World War II bunkers, pillboxes,
  artillery platforms, and
  observations towers

World War II ordnance

On the Pacific Coast of Colombia, the city of Buenaventura (population 320,000) used to dump its solid waste into Buenaventura Bay. With the right tide and wind conditions, this flotsam would float out of the bay and into the ocean, ending up on nearby barrier island beaches. During storms, the trash was carried landward, into the interior of the rain-forest-covered barrier islands. Ironically, buried trash items could then be used to estimate the age of storm-overwash sand deposits on the islands. We hasten to point out that most Colombian beaches on the Pacific are beautiful and pristine.

Shipwreck on a beach in Okhotsk, Siberia. Beaches in this vicinity have numerous shipwrecks, which in most cases are actually abandonments rather than wrecks. Photo courtesy of Nicolas Epanchin.

Myths and legends are easily created on beaches like the fog-shrouded Koekohe Beach, Otago, New Zealand, where the mysterious Moeraki Boulders might have been placed by giants. In fact, the boulders are resistant concretions that formed millions of years ago in the mudstone formation that is today being eroded by the Pacific's waves, leaving the rounded masses in the otherwise sandy beach. Photo courtesy of Brad Murray.

Latin beaches are not the only ones with a trash problem. In a single year the following items were found on a Massachusetts beach: refrigerators, shoes, lightbulbs, portable toilets, stepladders, tires, toys, balloons, batteries, fishing lines, crates, clothes, and much more.

Hats and caps are a common component of beach debris, most often blown off the heads of boaters and fishermen trolling just offshore. Shaving lotion, whiskey, vodka, gin, and beer bottles of all descriptions, brands, and nationalities are commonly found on beaches. If the bottles have resided on the beach for a few months, they may become frosted—abraded by wind-blown sand. Sometimes trash can be the nucleus for dune formation on the uppermost beach. Sand accumulates on the lee side of obstacles such as bottles and logs, and if vegetation is established, a dune may form.

Every few years, medical waste, including blood bags and syringes with needles, ends up on New Jersey beaches. In 2008, medical waste was found in a 50 mi (80 km)-long offshore garbage slick, with identifying labels indicating that the material came from two New York hospitals. Apparently the material, which was supposed to be burned in an incinerator, was hauled offshore by a contractor, who dumped it all close to the coast.

Imagine the change in the heart rate of the first beach strollers who spotted lead ingots on a South African Beach. Certainly these must have initially been mistaken for gold or silver ingots. The bars came ashore because of a strange quirk in the nature of storm waves. As anyone who has snorkeled near a surf zone can attest, breaking waves rock the water column back and forth. As it turns out, the forces that push bottom material in a landward direction are stronger than the forces that move it back toward the sea. Thus the lead ingots from a shipwreck were inched forward (in storms), centimeter by centimeter, until they reached the beach. The sand moved back and forth, too, but the ingots moved in only one direction, toward the beach, in response to the strongest currents.

Occasionally land animals find themselves on a beach where they don't want to be. Poisonous copperhead snakes are found (rarely) in the surf zone off some Outer Banks islands of North Carolina, probably washed to sea in a river flood. In South Africa, two years after the enactment of a prohibition on driving on almost all of the nation's beaches, leopards were occasionally spotted loping along the swash zone. Apparently when cars were still whizzing by, the big, shy cats avoided the beach, which was a normal part of their range.

A NATURAL LABORATORY

Beaches are a natural laboratory for the study of physical and biological processes and unique environments that have persisted through geologic time and left a rock record. Our fascination with beaches is what led us, four geologist authors, to write this book. In our combined one hundred plus years of studying beaches, we have each arrived at an appreciation of beaches that goes beyond the science. We have come to love the subject of our studies and to see the unique beauty in beaches. Besides the obvious exquisite experience of strolling on a beach, hearing the roar of the surf, and feeling the wind and the salt air, we have learned that a beach is a complicated creature that evolves in complex, yet predictable ways. We also have learned that the beach leaves all kinds of fascinating clues as to how it works, how it was formed and shaped; its surface is a record of timed events from the last season through the last week to today and the last few seconds. The beach is a natural mural waiting to be read and interpreted.

When you finish this book, you will know about beach characteristics ranging from the broadest scale of different types of coasts down to the individual beach features, including barking sand, heavy minerals, bubbly sand, the array of ripple marks, the significance of stained shells, and other interesting phenomena. In the end we hope that you too will feel the beauty in the beach as we do and that you will understand how it responds to tides, waves, and wind, how the critters that live in the beach manage to survive, and how humans can destroy the very features that we find unique.

Finally, we fear for the future of beaches. In this time of rising sea level, whether developed beaches survive or not will depend on how society answers the question, Which is more important, beaches or buildings?

2

BEACHES OF THE WORLD

WHAT IS A BEACH?

The beach is such a globally recognized feature that it hardly seems necessary to answer the question, What is a beach? We all have strong mental images of beaches lined with umbrellas, or beaches in the context of historical events, or that nearly universal, frequently happy memory of "my first trip to the beach." It is no surprise, then, that the definition of a beach varies among observers and even among scientists and engineers.

In the science of beaches, a biologist defines the beach by its principal natural biological function: "a habitat" for birds, turtles, shellfish, and meiofauna (microscopic organisms that live between the sand grains). Geologists and engineers define a beach as a deposit of sediment ranging in size from sand to boulders, formed by waves along a coastline. The beach is said to extend from the dune or vegetation line to an offshore depth at which the sediment is rarely moved and reworked by wave energy except during major storms. The underwater, offshore extension of the beach, the *shoreface*, commonly extends to a water depth of 30 to 65 ft (10 to 20 m) off sandy shores.

RECIPE FOR MAKING A BEACH

The prerequisites for a beach are simple: a supply of sand or gravel (sediment), the energy of the waves, a setting where sand can accumulate, and a definitive sea (or lake)

A beautiful, remote beach in Oman. The beach continues along the shoreline to the rocky cliffs in the background, which have very narrow beaches in front of them. Wide beaches such as this one are characteristic of fine-sand beaches. Photo courtesy of Miles Hayes.

level. Beaches form a dynamic equilibrium within these parameters. When one of the parameters changes, the others adjust accordingly.

Beach sediment comes in many sizes, shapes, colors, and compositions, and this variety gives rise to the great diversity in the appearances of the world's beaches (see chapter 3). The common stuff of beaches is sand and gravel; mud sediment is easily dispersed and transported away from the shore by breaking waves, ultimately settling to the seafloor in deeper offshore waters. Mud beaches (mud flats) are rare but arguably do occur on some low-wave-energy coasts such as the northeastern shores of South America, where enormous quantities of silt and clay from the Amazon River accumulate on very gently sloping coastlines.

Waves are needed to move beach sediment and shape it into the familiar (and perhaps unfamiliar) forms to which geologists have given various names (see chapters 4 and 5). Because waves differ in size, form, and other properties over time at any one beach, as well as all around the world, beach shapes vary with time and location (see chapter 4). This truth has given rise to our favorite sayings: "No two beaches are alike" and "Nothing on the beach stays the same."

For a beach to form, a coastal setting is required that is conducive to trapping sand or gravel, which allows for sediment accumulation. Natural irregularities or indentations in the coast are excellent locations for waves to form beaches because of the way waves are changed by headlands, embayments, and other topographic variations along the shoreline on the sea floor (see chapters 4 and 5). In contrast, shear, linear rock cliffs rarely host beaches.

Over a long period of time, waves tend to straighten out sandy shorelines. Crooked coastal-plain shores, formed when river valleys were flooded by the rising sea level at the end of the last ice advance, are straightened out by long chains of barrier islands and spits.

Finally, the elevation of sea level sets the geographic position of the beach, and in this time of a rising sea level, we will discuss the beach's ability to change position with sea-level change, particularly to transgress, or migrate landward with a rising sea level (see chapter 4).

CLASSIFICATION OF COASTS AND BEACHES

Just as the Inuit of the Arctic have many different names for snow because snow is important to them and behaves differently when it is wet or dry, warm or cold, old or

A crowded gravel beach at Albenga, Italy. Gravel beaches are generally difficult swimming platforms and are uncomfortable for lounge areas. This beach has the advantage of a railroad connection a few feet away, but this does not add a sense of peace and relaxation to the environment.

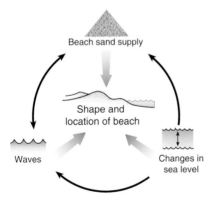

The beach is the product of sand supply and wave energy, and its position is controlled by sea level. The beach profile, or cross section, adjusts to these three controls and is said to be in equilibrium when it adjusts to these conditions. However, since wave energy and sand supply change, the beach profile is usually in a constant state of adjustment. In most parts of the world, the beach also must move landward as the sea level rises. Drawing by Charles Pilkey.

This red sandstone cliff in southern Portugal (with the Valle do Lobo resort development sitting on top) is an important source of sand for the adjacent beach as well as for the chain of seven barrier islands in the Portuguese Algarve. The cliff is easily erodible and has frequent small collapses. Though deaths on bluffed shorelines from collapses such as these are rare, a tourist lounging at the base of the cliff was killed by such a collapse.

A fan delta at the mouth of two intermittent streams in the desert of Oman. The importance of the streams as the source of sand for the beach is obvious. The lack of vegetation on the delta is a reflection of the arid climate. Photo provided by Miles Hayes courtesy of the Omani government.

young, so geologists have many different names for beaches, which are formed of different sands and gravels by waves that have variable characteristics and that display a variety of forms.

A fundamental aspect of natural science is the naming and classification of nature's features and processes. Nature, of course, does not necessarily adhere to our classifications, but still this is an important step by which we can understand the processes that shape an environment. In fact, a review of classifications of coasts over the last hundred years is partly an exercise in reviewing the history of our progress in understanding beaches.

One of the oldest coastal classifications was put forth by Columbia University geologist Douglas Johnson in his pioneering 1919 book, *Shore Processes and Shoreline Development*. Now long outdated, his classification held sway among coastal scientists for many years. He recognized four major types of coasts. The first was a coast that showed signs of being recently submerged; examples include the coasts along the drowned river valleys that make up the Maine coast, Chesapeake and Delaware bays, Spencer Gulf in Australia, and the fjords of Norway. Johnson's second type of coast was one that, he incorrectly believed, had recently emerged from the sea. In this category he included coastal-plain shorelines such as the barrier islands and marshes of southern New Jersey and the Dutch-German Frisian Islands. He described a third shoreline type as neutral with respect to sea level, depending on neither submergence nor emergence. This category included deltas (which are in fact among the most rapidly drowning shorelines in the world), coral reefs (which are closely linked to changing sea level), and mid-ocean volcanoes.

Finally, Johnson's compound shorelines displayed evidence of both submergence and emergence. As examples, he cited North Carolina's Outer Banks, which he believed was recently emergent, and the coastline of Pamlico Sound, which he thought had earlier drowned. As it turns out, however, most of the world's coasts fall into the submerged category, reflecting the fact that sea level rose to its present level in just the last few thousand years, after the glaciers melted. Johnson's classification, although incorrect on its fundamental points, was an important start because it recognized that fluctuations of sea level, both local and global, play a major role in determining the character of beaches and coasts.

By the 1930s, a better understanding of coasts and beaches was taking shape, and new classifications evolved. Francis Shepard divided coasts into two types, primary and secondary. Primary coasts are those that owe their shape mainly to processes that occur on land. These include alluvial fan coasts such as those in front of the glaciers of southeast Iceland, lava coasts such as those in Hawaii, fjord (glaciated) coasts, as in Norway, and river-formed coasts, as on the deltas of the Niger, Mississippi, Nile, and Ganges rivers.

Shepard's secondary coasts are formed by marine processes; they include the wave-straightened coasts of North Africa and Siberia; the barrier island shoreline of the United States' East and Gulf coasts, Brazil, and China; and the Arctic barrier islands of the North Slope of Alaska. He also included the coral reef coasts of the Pacific and Indian oceans' atolls and various Caribbean Islands in his secondary division.

Shepard, who was at Scripps Institution of Oceanography in California at the time, made an even more important observation, though on a smaller scale. He noticed that there was a difference between winter and summer beaches as well as between beaches before and after storms. He noted that the beaches around La Jolla, California, were narrow and sometimes covered with gravel in winter, when major storms occur, but in the summer the same beaches were sandy and much wider. For years this summer-winter difference was assumed to be a global phenomenon, but as it turns out, it is really a matter of non-storm- and storm-dominated shorelines; beaches everywhere have their own climate and character.

Armstrong Price was a remarkable Texas A&M geologist whose career spanned the years from the 1920s to 1983, when he presented his last technical paper at a professional meeting, at age ninety-nine. Price was a contemporary of Francis Shepard's, but he formulated a beach classification influenced largely by his extensive studies along the Mexican and U.S. shores of the Gulf of Mexico. This classification is still widely used, albeit with some modification, and subdivides beaches according to wave energy as measured by the typical wave height. Price recognized three classes of beaches: low-wave-energy beaches (characterized by waves 1 ft [30 cm] high), moderate-wave-energy beaches (characterized by waves 2 ft [60 cm] high), and high-wave-energy beaches (characterized by waves greater than 2 ft [60 cm] high). On most other open-ocean shorelines (e.g., the east and west coasts of the Americas), the energy subdivisions would involve much greater wave heights. Price also identified a zero-wave-energy category, which applied only to Florida's west coast, along the northeast corner of the Gulf of Mexico.

In addition, Price observed that the wave height increased from east to west along the northern coast of the Gulf of Mexico, which he attributed to the fact that the conti-

The old stumps protruding from this beach surface at Nags Head, North Carolina, are remnants of an old maritime forest, buried over time and now reexposed as the shoreline retreats landward with the sea-level rise and barrier island migration. The stumps are usually exposed after minor storms from the northeast. Photo courtesy of Ray Midgett.

A view of the same beach two days later. In the interim, there was a minor storm with winds from the southeast that pushed sand ashore on this north-south-trending beach and reburied the stumps. Photo courtesy of Ray Midgett.

nental shelf got narrower and steeper in that direction. This relationship turns out to be a universal truth for beaches. Other things being equal (such as storm frequency and *fetch*, the distance over which winds can blow to form the waves that ultimately strike a beach), the narrower and steeper the continental shelf, the higher the waves striking the beach. On gently sloping, wide continental shelves, the waves lose much of their energy as they drag bottom across wide areas of shallow shelf. However, the greater the fetch, the bigger the waves will be. The beaches of the U.S. states of Georgia and North Carolina face fetches of thousands of miles of open ocean, but Cape Hatteras, North Carolina, is a high-wave-energy beach because the shelf width is only around 30 mi (about 50 km), whereas the 80 mi (about 130 km)-wide continental shelf off Georgia is responsible for the low-wave-energy beaches there.

In 1952, New Zealand geologist C.A. Cotton introduced the idea of using regional stability as the primary category under which to subdivide coasts. Unstable or mobile coasts are those experiencing mountain building and occasional earthquakes. Mobile coasts include the west coasts of North and South America, and stable coasts include the east coasts of North and South America and all coasts of Africa. Coastal instability (mountain building) is important because it is responsible for different sediment types, wave climates, and other beach characteristics, as discussed later in this chapter.

Scientists in the 1960s and 1970s continued to refine and build on the works of these pioneers. From his base in Australia, J.L. Davies enlarged on the Armstrong Price approach and came up with a widely used so-called dynamic classification of coasts. The basis of his classification is the types of waves, broadly defined, that typically strike the beaches on a particular coast. These are the waves that are responsible for most of the changes in beaches in a particular region. He recognized six types of *wave climates* that affect beaches around the world. Following are the types of wave climates and their usual locations:

- *Waves from frequent storms*—southern tip of South America, New England and eastern Canada, southern Greenland, Europe north of Spain, and Pacific Russia and Alaska
- *West-coast swells*—West Africa, India, Southeast Asia, Western Australia, and much of the western Americas
- *East-coast swells*—East Africa, east Australia, Brazil, and east New Zealand
- *Trade- and monsoon-wind waves*—Southwest Australia, Madagascar, northern Brazil, and the outer Caribbean
- *Tropical cyclones (hurricanes)*—southeast coast of the United States, Northwest Australia, China, and Japan
- *Low-energy wave environments*—the Arctic, bays, and estuaries

A low-wave-energy Antarctic cobble beach. The large size of the beach materials most likely is due not to high wave energy, but rather to the constant flow of summer meltwater from the adjacent glacier across the beach. The penguins are shown for scale. Photo courtesy of Norma Longo.

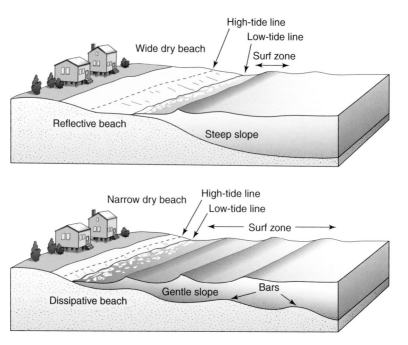

Diagrams comparing reflective and dissipative beaches show the wide, reflective beach with a gentler offshore profile in contrast to the narrower, dissipative beach with offshore sandbars. Drawing by Charles Pilkey.

Clearly Davis's global view captured a lot of the variability found around the world. Different climates produce differing wind regimes; differing continental orientations and geometries likewise interact in variable ways with waves.

In the 1980s, Donn Wright and Andy Short, leaning on Shepard's earlier thinking, proposed another way to describe beaches. They recognized two endpoint-type beaches with several intermediate types between them. The endpoints are *dissipative beaches* and *reflective beaches*. This classification of beaches was developed primarily in Australia, but it has been applied all around the world with some success. The factors that determine the beach type include only the wave height and size of the beach sediment.

Dissipative beaches are wide, flat, and generally fine grained. Because of their width, waves dissipate their energy as they cross the outer beach. The waves that finally reach the swash zone tend to grow in height and are popular with the recreation-oriented public. *Offshore bars*, the locations of which are marked by the long white lines of breaking waves offshore, are well developed on dissipative beaches. On some dissipative beaches, as many as five or six lines of bars marked by breaking waves can be spotted, but there are more commonly two or three bars. Separating the bars (as every swimmer knows) are troughs that can be 3 to 6 ft (1 to 2 m) deep.

Reflective beaches are steep, often have coarser sand grains or gravel, and have no offshore bars. Thus, the waves lose relatively little energy until they actually break right on the beach. Reflective beaches typically have small breaking waves that sometimes are reflected back off the beach. Often these beaches have beach cusps (see chapter 5). In general, reflective beaches also have a wider dry beach than dissipative beaches; however, some eroding beaches tend to be reflective and have no dry beach. The *dry beach* is defined as the stretch of beach that remains for us to play on at high tide, the zone between the high-tide line and the first dunes or vegetation.

CLASSIFICATIONS AND A GLOBAL MODEL

Clearly, coastal workers' thinking is strongly influenced by the coasts and beaches that they study, and it takes global experience or a global model to produce a universal classification. From the 1960s through the late twentieth century, the global model known as *plate tectonics* grew out of the earlier hypothesis of continental drift and changed almost every aspect of earth science. The C.A. Cotton coastal-stability approach anticipated this and eventually led

Pearl Beach in New South Wales, Australia, is an example of a reflective beach. The relatively small waves break right on the beach. Spectacular rip currents can be seen spaced at regular intervals controlled by beach cusps, which are well developed on this beach. Photo courtesy of Andy Short.

Dog Fence Beach in South Australia is an example of a dissipative beach, characterized by a wide, shallow platform of sand in front of the beach. Sandbars on this platform cause the bigger waves to break well offshore, so the waves that finally make it to the beach are relatively small. In this photo, there are at least three sandbars and possibly more. The wave patterns are made indistinct because of an offshore wind blowing parts of the crests seaward. Photo courtesy of Andy Short.

to the 1971 coastal classification by Doug Inman and C.E. Nordstrom of Scripps Institution of Oceanography, which was based on this revolutionary new theory of plate

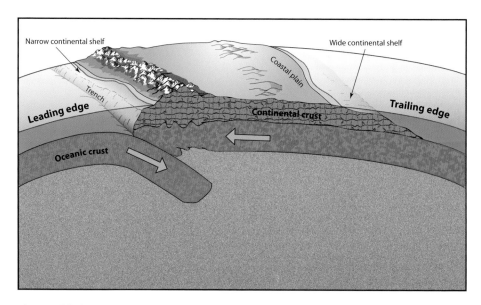

Narrow continental shelf

Wide continental shelf

Coastal plain

Trench

Leading edge

Continental crust

Trailing edge

Oceanic crust

This simplified plate tectonics model shows a collision-edge coast on the left, where subduction of ocean crust under the continent causes volcanism, deformation, and uplift of mountains, resulting in a narrow continental shelf, and rugged, rocky coasts with pocket beaches. In contrast, the trailing-edge coast on the right is bordered by a low coastal plain and a wide continental shelf. Here, conditions are more favorable to continuous mainland beaches and barrier island chains. Drawing by Charles Pilkey.

tectonics. In this approach, beaches and coasts are categorized according to their plate tectonic setting.

Plates are the rigid upper parts of the Earth's surface that move about. As the eight major and numerous smaller plates shift, they either collide with one another and end the life of an ocean; move apart, creating new ocean basin; or slide past each other (a type of major faulting). When a plate incorporating a continent collides into an oceanic plate, rocks on the edge of the continent are deformed and uplifted to form a coastal mountain range such as the South American Andes, the North American Pacific Coastal Ranges, and large, mountainous islands such as Japan and the Philippines, as well as mountainous peninsulas such as Italy and Indonesia. This continental plate deformation and associated volcanic activity result from the oceanic plate moving under (subducting) the continental plate. From the perspective of the shoreline, the results are *collision* or *leading-edge coasts*.

Where coastal mountains meet the sea, spectacular sea cliffs may occur. The beaches (called *pocket beaches*) that exist in these locations often occur in coves, are typically small, and are surrounded by the cliffs. Commonly on leading-edge coasts, the land is either rising or falling. For example, in California, Peru, and Chile, upraised shorelines and marine terraces can be seen landward of today's beaches, like giant steps going up to the mountain front. Some of these uplifted ancient shorelines in South America have important associated archeological sites of early coastal cultures. In the very large 1964 Good Friday Earthquake in Alaska, some areas near Kodiak were permanently raised by 30 ft (9.1 m).

In contrast, in that same 1964 earthquake, east of Anchorage, Alaska, areas around the head of Turnagain Arm near Portage dropped 8 ft (2.4 m), requiring reconstruction and fill to raise the Seward Highway above the new high-tide mark. Beaches of the Kenai Peninsula sank 2 to 3 ft (1 m), causing wide beaches to become narrow beaches overnight. Along the Pacific Coast of Colombia, South America, numerous earthquakes, associated with subduction along the plate boundary, have caused local overnight shoreline drops of several feet or so, sometimes generating small but deadly tsunamis.

Mid-ocean volcanic islands also are related to plate tectonics, forming either where plates are moving apart (seafloor spreading) or where hot spots exist below the oceanic plate; as the plate moves over the hot spot, a chain of volcanoes evolves. Iceland, the Azores, and the Canary Islands are Atlantic examples, and the various Pacific island chains such as the Hawaiian Islands are typical. These oceanic volcanic landmasses have shorelines with pocket beaches, small spits, and very rarely barrier islands.

The situation opposite the one that characterizes the edges of colliding plates is found on the other side of a moving continental plate, that is, the *trailing-edge coast*. When a continent splits apart and the two formerly joined landmasses move away from each other, a new ocean basin forms. The Atlantic basin formed this way when Europe and North America rifted apart (about 250 million years ago), and similarly, ocean basins formed when Africa and South America separated and when Antarctica and Australia separated. The east coasts of North and South America and the west coasts of Europe and Africa are on the trailing edges of these moving continents.

As the continents spread apart, they shed huge quantities of sediment into the widening sea, building up a sedimentary wedge thousands of feet thick known as a *coastal plain*. Such coastal plains have very gentle seaward-dipping slopes. In almost all cases, coastal plains are characterized by wide continental shelves (30 to 100 mi [50 to 160 km]), resulting in relatively low-wave-energy beaches. In South America, the coastal-plain setting favors long, continuous beaches on the mainland shore. In North America, a similar setting has resulted in lengthy chains of barrier islands.

BEACH SEDIMENTS AND THE PLATE TECTONIC SETTING

The contrasting plate tectonic settings even help to explain some broad aspects of beach sediment composition. Volcanic islands bring to mind black or green sands, sands that are usually dark in color because they are derived from volcanic rocks. The exceptions are where reefs fringe the islands and light-colored calcareous sands wash from the reefs to the shore, producing light-colored beaches.

Upper left A typical mountainous leading-edge coast in the Big Sur section of California. Here the beaches are narrow to nonexistent and are mostly restricted to pocket beaches in the narrow embayments (right foreground, the beach just out of the photo). A significant fraction of beach sediment on leading-edge coasts is derived from erosion of adjacent cliffs and bluffs. Where rivers are present, they often furnish sand directly to beaches during floods. If such rivers are dammed, sand is trapped behind the dam and erosion commences on the beaches. Photo courtesy of Miles Hayes.

Lower left Pocket beaches also result from man-made coastal structures, which behave as headlands, as in this beach in Dubai, yoked by jetties and riprap walls. The beaches here are highly manipulated and shaped by humans, but they can still be beautiful places for swimming and relaxing, as demonstrated by the man sunbathing in the foreground.

Upper right Pocket beaches, like this one north of Sydney, Australia, typically are confined between rocky headlands. Erosional features are dominant, such as the wave-cut cliffs on the headlands and the back of the beach. A wave-cut terrace in the foreground has exposed well-developed joints, seen in the rock surface at the front of the photo. When such rocks are subjected to high storm waves, the joints or cracks in the rock facilitate its breakup.

Lower right A trailing-edge coast on Prince Edward Island, Canada, far from any mountain range. This beach has a sand supply, as indicated by the extensive array of sand waves in front of the beach; however, the beach is not particularly wide, and an erosional sea cliff is apparent along part of the shore. Photo courtesy of Miles Hayes.

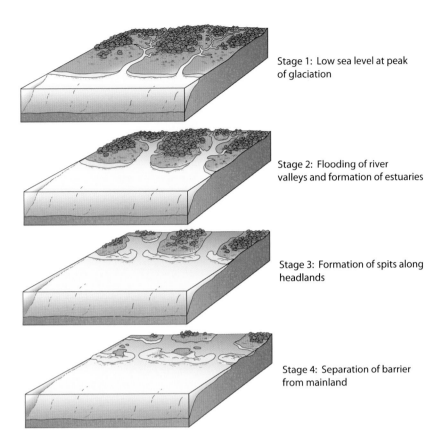

Stage 1: Low sea level at peak of glaciation

Stage 2: Flooding of river valleys and formation of estuaries

Stage 3: Formation of spits along headlands

Stage 4: Separation of barrier from mainland

Diagram of a common series of events that gave rise to barrier islands after the Ice Age. When water was locked up in extensive glaciers, sea level was lower. As the glaciers melted and sea level rose, river mouths were flooded, forming estuaries separated by headlands. The headlands provided anchors from which spits could grow into the mouths of the estuaries, until storms truncated the spits into separate islands by forming inlets. Once a barrier island forms, it will migrate landward if the sea level continues to rise. Drawing by Charles Pilkey.

Both continental leading-edge and volcanic island beaches tend to have young, or "immature," sediments. The sand is described as immature because it has not traveled long distances or undergone abrasion and atmospheric weathering for long periods of geologic time. The mountain range or volcanic sediment source is often visible from the beach, indicating the short distance of transport and the rapid erosion due to the steep mountain slopes. The sands are often dark colored because they have a large variety of mineral types in them, plus sand-size fragments of dark-colored rocks. The principal sources of beach sands on leading-edge coasts are local streams, rivers, landslides, and slumps. Leading-edge coasts of continents as well as coasts of volcanic islands tend to have narrow continental shelves. As a consequence, the beaches in these locations are usually medium- to high-wave-energy beaches, according to the Armstrong Price classification.

In contrast, the sediments delivered to trailing-edge beaches are mature. The sands are more likely to be "recycled" from earlier sedimentary rocks and may also have been reworked, having suffered a long abrasion history, which tends to eliminate less stable minerals and the sand-size fragments of fine-grained rocks that one might find on a leading-edge coast. As a consequence, coastal-plain beaches tend to be dominated by relatively stable minerals—a residuum of hard quartz and feldspar grains—plus shell fragments. In a very general way, trailing-edge beaches tend to be lighter colored and finer grained than the beaches on leading-edge coasts. The main source of beach sand on trailing edges is the continental shelf, from where waves push the grains ashore. River sands on such coasts tend to be deposited at the upper ends of the estuaries, far from the coast, and do not make it to the beaches.

All generalizations, however, have their exceptions. Glaciated coasts, even on trailing edges, are characterized by immature sediments because the sediments are derived from a wide variety of sources, often including crystalline igneous and metamorphic rocks such as granites and gneisses. The materials have sometimes been transported for long distances, but on the conveyor belt of the glacier, without chemical weathering. Glacial tills are a homogenized mixture of all of the rocks over which the glacier moved, usually with the entire range of sizes from boulders to clay. The rocky beaches of the Scandinavian countries, the British Isles, eastern Canada, and New England are good examples of reworked, glacially derived sediments on a trailing-edge coast.

BEACHES AS LANDFORMS

Another approach useful to understanding beaches is to class them in terms of the coastal landform that the beach shapes or on which it is located. Common types include the following:

- *Barrier island beaches:* Barrier islands are generally sandy islands that are longer than they are wide and that lie parallel to the coast. Typically backed by salt marshes, lagoons, or both, barrier islands are found along all continents except Antarctica. The beaches of these barriers are highly favored as tourist sites.

- *Mainland beaches:* Mainland beaches are those beaches that directly abut the land and usually receive their sediment supply from streams, from episodic erosion of the cliff or bluff behind the beach, or simply from reworked shelf materials. Virtually all beaches on the edges of continental landmasses are mainland beaches, but subcategories may be recognized. Trailing-edge coasts often have beaches that are quite long (many miles) and straight,

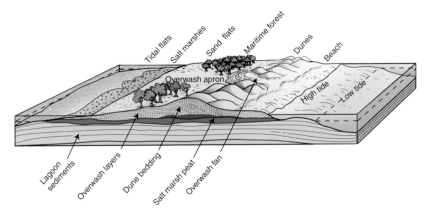

This diagram of a barrier island shows the dominant environments and is indicative of how an island can migrate. Sand is carried across the island by storm waves as overwash, by landward dune migration, and through the inlets at the ends of the island (not shown). The sand is carried to the back of the island and into the marsh, or, in the case of the inlet, onto the flood tidal delta. These deposits form the platform on which the island migrates. In the process, what once existed on the back of the island will first be buried, and then ultimately exposed again on the front of the island (resulting in, e.g., maritime tree stumps and marsh muds on the beach). Drawing by Charles Pilkey.

whereas leading-edge coasts are likely to have pocket beaches, as noted earlier.

· *Pocket beaches:* A pocket beach is a beach tucked between rocky headlands. Most pocket beaches are relatively short in length, as the name implies, and may be backed by a marsh, a bay, or a cliff. Often these beaches afford the only recreational swimming areas along rocky shorelines.

· *Cuspate forelands and capes*: These triangular depositional areas extend seaward to a rounded point and are fringed by sand or gravel beaches. In Great Britain, a cuspate foreland is termed a *ness* (e.g., Dungeness). Capes also are seaward extensions of the coast and are often due to rocky headlands; however, along sandy coasts, capes are often low-lying areas of accreting beaches, as on barrier island chains (e.g., Cape Canaveral, Florida). The different orientations of the adjacent beaches are a consequence of or they can alter the direction of longshore sediment transport.

· *Smaller beach landforms:* Several beach landforms have unique origins or characteristics that give them name recognition. *Spits* are beach-bar complexes that grow out from the coast into open water, so that one end is attached to land and the other growing end terminates in open water, usually in a tidal inlet or embayment. Somewhat similar in form to barrier islands,

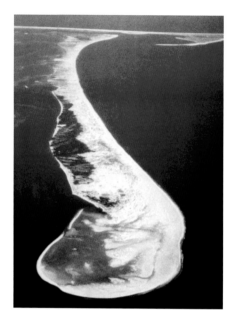

Left This aerial view provides a good example of a barrier island (Shackleford Banks, North Carolina). At the top of the photo is Cape Lookout, a cuspate foreland, separated by an inlet from Shackleford. The dark patch on the back of the island is a maritime forest, separated from the beach by a wide zone of sand dunes. Sand carried along by longshore drift is building out into the inlet in the foreground of the photo. Barrier islands with their accompanying sandy or gravelly beaches make up close to 12 percent of the length of the world's open ocean shorelines.

Upper right The west end of Dauphin Island, Alabama, is perhaps the most hazardous barrier island development in the United States. It has been destroyed by hurricanes several times but always redeveloped. This was the scene shortly after Hurricane Katrina in 2005. Note the widening that occurred as sand washed from the front of the island to the bay side as overwash fans. The houses on the bay

side of the island resided next to the water before Katrina. At the end of the spit in the foreground of the photo, the sand is moving from right to left. The beach in front of the island has been nourished multiple times. The lesson is that beaches and islands migrate; houses do not, unless you count floating. Photo courtesy of the Alabama Department of Public Safety.

Lower right A South Carolina spit extending to the south from Kiawah Island near the city of Charleston. Captain Sam's Inlet, in the foreground, is migrating to the south as the spit grows, forming new land. Plans are afoot to develop this very low island segment, a most dangerous location for buildings because it is too low and will be cut off when the inlet reestablishes its former position during a storm. In the past, storms have cut new inlets across the narrow part of the spit (note the narrow neck in the background).

spits depend on longshore transport from the adjacent shoreline for their sand supply. Spits may grow all the way across the mouth of an embayment to form a beach barrier known as a baymouth bar. Where spits grow in two directions off of a headland, a winged spit forms. Spitlike features may grow off of a coastline and connect the mainland to an island by a narrow beach neck, producing a *tombolo*.

Similar to a baymouth bar, sandbars can form completely across the mouths of small rivers. When floods occur, the bar is broken and for a while an inlet remains open, only to be closed when river flow is again low and the bar re-forms. This is a common occurrence along mountainous coasts with small rivers.

CLIMATE AND TYPES OF COASTS

In addition to the historic classifications of coasts, climate and past climate history certainly are recognized as controlling factors in types of beaches. The more common of these climatic shore types include the following:

Glaciated coasts: These coasts bound land that was once occupied by glaciers, such as Scandinavia, Ireland, New England, and Canada. Beaches derive their sediment from deposits of sand and gravel left behind by the glaciers.

Glacier coasts: Coastlines with nearby active glaciers are subject to seasonal sediment-laden, freshwater surges (e.g., the *jökulhlaup* of southeast Iceland), resulting in very dynamic, sediment-rich beaches that usually have a high gravel content.

Arid coasts: Arid coasts are characterized by a lack of sand-stabilizing vegetation, which allows large dune fields to be formed as sand is exchanged back and forth between the beach and the hinterland (typical locations are Namibia, Africa, and northern Peru). In extremely arid cases, on very flat land adjacent to the ocean, the land may be flooded by extreme high spring tides or during storms with strong onshore winds. The floodwaters then evaporate off these *supratidal flats (sabkhas)* causing the precipitation of minerals (e.g., the Abu Dhabi salt deposits of the Persian Gulf sabkhas). The supratidal flat itself, above the normal high tide line, may not be sandy, but a fringing sandy beach with small dunes may be present.

Tropical coasts: The nature of beaches in the tropics depends on the geologic setting and ranges from dark sand and gravel near mountain ranges (e.g., the Pacific Coast of Colombia) to mud near river mouths (e.g., Suriname, South America) to white sand onshore of coral reefs (e.g., Kenya, Africa).

Arctic coasts: Beaches in the Arctic are undergoing rapid changes because of melting permafrost (within the beach sand) and lengthening periods of

Upper left A glaciated coast formed when glaciers once extended beyond this Alaskan shore, but recent processes have modified the coast by filling in a former embayment (beach ridges to the right of the stream) and straightening the shoreline with incorporated barrier beaches and spits. The small delta is evidence that the stream is an important sediment source for the adjacent beach. Photo courtesy of Miles Hayes.

Lower left This tropical barrier island (Santa Barbara Island) is on the west coast of Colombia, near the equator. The beaches have a low carbonate content, in spite of the warm water, because of the high noncarbonate sand supply coming from the nearby Andes, which dilutes the carbonate fraction. The tropical rainforest extends right up to the beach.

Upper right Arid conditions can prevail in cold climates such as in the Straits of Magellan in Chile, the home of this long gravel beach. Photo courtesy of Miles Hayes.

Lower right A spit at Point Barrow, Alaska, enclosed in spring ice off the North Slope of Alaska. Arctic barrier island and spit beaches now are retreating at rapid rates because global climate change is melting the permafrost and "loosening" beach sand during the summer. In addition, the arctic sea ice is shrinking, which means that ice-free conditions, with waves and storm activity, now occur for longer periods of time each year. Photo courtesy of Miles Hayes.

THE WORLD'S LONGEST BEACH

Travel brochures often tout locales with miles and miles or tens of kilometers of continuous beach, appealing to a common perception that the perfect beach ought to stretch out to the horizon. Competitive claims thus arise as to the longest, widest, softest, firmest, whitest, and yes, sexiest beach in the world. For instance, a sign that says "World's Longest Beach" greets visitors to the town of Long Beach, Washington, but at best the beach's 23 to 28 mi (37 to 45 km) is not even close to world-class length. It turns out that the sign is really referring to the world's longest drivable beach, but this claim is not supportable, either, because much of the world's longest barrier island, Padre Island, Texas, is drivable. Padre Island's length is given as 130 mi (209 km), though the artificial Mansfield Cut interrupts that distance.

Although the Long Beach claim has been modified again, now assigning it the title of the world's longest continuous peninsular beach, that is probably debatable, too. The Arabat Spit, in the Sea of Azov, between Ukraine and Crimea, is approximately 68 mi (110 km) long and probably beats out Long Beach for drivable miles. Whatever its claim, Long Beach is still a long, fine beach and is certainly worth a visit!

Of course, Long Beach is not the only beach to make an exaggerated claim to lure tourists. Marina Beach, India, on the Bay of Bengal, makes a similar claim but is not even close to being the world's longest beach. Perhaps the Indian claim is because of its competition with the neighboring Bangladeshi beach that was long held to be the longest continuous beach in the world: Cox's Bazar, usually listed at 150 mi (more than 240 km). More recently, however, Praia do Cassino, in Rio Grande do Sol, Brazil, has generally been agreed to be the longest beach, at about 158 mi (254 km).

Australia boasts some world-class contenders for uninterrupted beach length. A.D. Short and C.D. Woodroffe note in their 2009 book, *The Coast of Australia*, that Coorong-Middleton Beach, in South Australia, is more than 120 mi (194 km) of continuous beach. The famous 90 Mile Beach in the southeastern part of Victoria is about 94 mi (151 km) long, and 80 Mile Beach in West Australia is about 78 mi (125 km) long. New Zealand also has a 90 Mile Beach, on North Island, but this misnomer arose from an error in the early estimates of the actual beach length, which is given as 55 to 64 mi (88 to 103 km), certainly not 90 mi (145 km), but more than long enough for the annual marathon held there.

Not to be outdone, Great Lake communities make claims for longest freshwater beach length: Wasaga Beach, Ontario, Canada, claims to be the longest, at about 8 mi (13 km). However, we believe that during times of low lake level in the Great Lakes, some other beaches exceed that length.

Now that Google Earth is available, we expect that someone will set out to measure some long stretches of beach in nontourist areas such as Somalia, Libya, and Namibia to challenge the Brazilian superbeach!

ice-free ocean expanses due to Arctic warming (e.g., in Siberia, northern Canada, and the North Slope of Alaska). The longer periods of ice-free ocean lead to longer periods of waves pounding the beaches, the sand on which is no longer cemented by permafrost.

Delta coasts: Although the location of large deltas is independent of climate, the type of vegetation that dominates the emergent surface of the delta *is* a function of climate (e.g., tropical mangroves are excellent sediment trappers, aiding in the formation of the delta). The sediment type on delta beaches will vary depending on the river drainage areas. Such sources range from active glaciers providing sediment ranging from mud to gravel (e.g., the Copper River, Alaska), to major rivers providing mud and sand (e.g., the Mississippi River), to rivers providing only mud (e.g., the Amazon River). Most, if not all, large deltas are sinking rapidly.

WE STAND ON THEIR SHOULDERS

There is more to beaches than their classification. In the chapters that follow, the discussions of minor features in the surface sands of beaches owe much to the observations of outstanding coastal scientists who went before us. These folks had enough intellectual curiosity to get down on their knees and closely examine the unique aspects of the beach. They were the scientists who dug long trenches across the beach to see how beaches work, who visited the beaches immediately after storms and during all seasons of the year. Nobody particularly suffers from having to spend a lot of time on beaches, but for some it was almost all work and little play. At work, these scientists were probably easily identifiable as the guys strolling down the beach wearing long pants! Today, even the most casual student of beaches stands on the shoulders of those who were curious enough to ask what, why, how, and when in regard to all aspects of beaches.

Everybody loves beaches. Who would not want to study them! A number of prominent scientists who developed our understanding of beach evolution are recognized here. Like a best-beach list, this one, too, is incomplete, but it reflects some of the most prominent scientists and demonstrates that they comprise a very international group.

- Vsevolod Zenkovich of the Oceanology Institute of the USSR Academy of Sciences first published *Coastal Dynamics and Morphology* in 1946.
- Cuchlaine King of the University of Nottingham published *Beaches and Coasts* in 1975 (second edition), considered the first book covering broad aspects of coastal geology.
- Willard Bascom, founder of Ocean Science and Engineering, a Southern California company, popularized beach science in his 1964 book *Waves and Beaches*.
- J.R.L. Allen of Reading University, England, presented field descriptions as well as the physics of the origin of sedimentary structures in his classic 1982 book, *Sedimentary Structures: Their Character and Physical Basis*.
- Hans Reineck of the Senckenberg Research Station for Marine Geology, Wilhelmshaven, Germany, and I.B. Singh presented an excellent summary of beach surface features in their 1975 book, *Depositional Sedimentary Environments*.
- K.O. Emery, who studied beaches extensively while at the University of Southern California, summarized his views of beaches in *The Sea off Southern California*, published in 1960.
- E.C.F. Bird, an Australian geologist, wrote a number of books about coasts, the most pertinent of which perhaps is *Coasts: An Introduction to Coastal Geomorphology*, published in 1984.
- Miles O. Hayes, formerly of the University of South Carolina, starting in 1967 published a number of field trip guide books and other publications, bringing a new understanding of the third dimension of beaches, as well as an important classification of barrier islands.
- Andy Short of the University of Sydney published several books on coastal hazards in Australia, devised the dissipative-reflective classification, and summed it all up in his 2009 book with C.D. Woodroffe, *The Coast of Australia*.

SOCIETAL CLASSIFICATIONS OF BEACHES

Lists of the "best" beaches abound on the Internet, on television, and in tourist literature. The rankings, which can be based on detailed lists of beach characteristics or (very often) on the whims of the tourist industry or chambers of commerce, include the best beaches for lovers, solitude, gays, nudists, surfers, small children, bird-watching, snorkeling, and so on. The criteria used to determine the "best" beaches sometimes have more to do with hotels, restaurants, amusement parks, and other nonbeach attractions than with the beaches themselves.

Following is a list of the world's best beaches—or at least a list of beaches that someone or some organization in each country has claimed to be a great one, among the best in the country. Ranking beaches is a dubious exercise: Not all beach lovers are looking for the same beach qualities. Someone interested in solitude would rank beaches quite differently from someone interested primarily in nightlife, and surfers inhabit shores that snorkelers do not. If nothing else, however, the list shows that love of beaches, whether for their economic value or their ability to enhance a community's quality of life, is a global phenomenon.

Abu Dhabi, U.A.E.: Emirates Palace
  Beach
Albania: Viora Beach
Algeria: Palm Beach, Sidi Fredj
Andaman Islands: Beach #7,
  Havelock Island
Angola: Amelia Beach
Argentina: Mar del Plata
Aruba: Palm Beach
Australia: Bondi Beach, Sydney;
  Surfers Paradise; Whitehaven
  Beach
Azerbaijan: Shikhov Beach
Bahamas: Pink Sands Beach
Bangladesh: Cox's Bazar
Barbados: Bottom Bay
Belgium: Vosseslag, Mispelburg
Belize: Ambergris Caye
Benin: Cotonou Beach
Bermuda: Pink Sands Beach

Bora Bora: Matira Beach
Brazil: Ipanema Beach, Rio
  de Janeiro; Porto da Barra,
  Salvador; Porto de Galinhas
  Beach; Copacabana Beach
Brunei: Muara Beach
Bulgaria: Balchik
Cambodia: Sihanoukville
Canada: Qualicum Beach, British
  Columbia; Sandy Cove, Nova Scotia
Cape Verde: Santa Monica, Boa Vista
Cayman Islands: Seven Mile Beach
Chile: Easter Island
China: Huiquan Beach, Qingdao
Cook Islands: Aroa, Aitutaki
Colombia: Cartagena
Costa Rica: Playa Grande, Guanacaste
Croatia: Golden Horn Beach, Brač
Cuba: Varadero Beach
Denmark: Lokken Beach

Dominica: Champagne Beach
Dominican Republic: Playa Grande,
    Rio San Juan
Dubai, U.A.E.: Jumeirah Beach
Ecuador: Atacames
Egypt: Laguna, Bay of Dahab;
    Hurghada Beach, Ahmar
El Salvador: Playa el Zonte, La
    Libertad
Eritrea: Assab Beach
Estonia: Pärnu Beach
Falkland Islands: Bertha's Beach
Fiji Islands: Natadola Beach,
    Viti Levu
Finland: Yyteri Beach, Pori
France: Conche des Baleines, Île de
    Ré; Plage de Tahiti, French
    Riviera; Rondinara Bay, Corsica
French Guyana: Kourou Beach
French Polynesia: Tikehau, Rangiroa
Gabon: Ekwata Beach, Libreville
Galapagos: Gardner Bay
Gambia: Bijilo Beach
Germany: Kampen, Sylt
Gibraltar: Catalan Bay
Greece: Corfu; Paradise Beach,
    Mykonos
Grenada: Anse La Roche
Guadeloupe: Plage de Malendure
Guam: Haputo Beach
Guatemala: Monterrico
Guyana: Shell Beach
Haiti: Labadee, Jacmel
Honduras: West Bay Beach, Roatan
Iceland: Nauthólsvík Beach,
    Reykjavík
India: Arambol Beach, Goa; Juhu
    Beach, Mumbai; Marina Beach,
    Chennai

Indonesia: Seminyak, Bali
Iran: Kish Island
Ireland: Dogs Bay, Roundstone,
    Galway
Israel: Gordon Beach, Tel Aviv
Italy: Latina, Sperlonga Beach;
    Salerno, Pollica; Sardinia, Cala
    Luna; Sicily, Marina di Ragusa
Ivory Coast: Bassam Beach
Jamaica: Doctor's Cave, Montego Bay
Japan: Kujukuri Beach, Boso;
    Shimoda, Izu; Shonan Coast;
    Miyakojima, Okinawa
Jordan: Aqaba
Kenya: Lamu Beach
Kuala Lumpur: Emerald Bay Beach
Kuwait: Messila Beach
Latvia: Jūrmala
Lebanon: Bamboo Bay
Liberia: Bernard's Beach
Lithuania: Palanga
Libya: Farwa Island Beach
Madagascar: Île aux Nattes
Malaysia: Datai Beach, Langkawi
Maldives: Huvafen Fushi
Malta: Ramia Bay, Gozo
Martinique: Le Diamant
Marshall Islands: Laura Beach,
    Majuro
Mauritania: Mauritania Beach
Mauritius: St. Gerán
Mexico: Acapulco, Guerrero; Mayan
    Riviera, Yucatán
Monaco: Larvotto Beach
Montenegro: Ladies Beach, Ulcinj
Morocco: Agadir Beach
Mozambique: Bazaruto Island
Myanmar: Ngwe Saung Beach
Namibia: Irgendwo Beach

Netherlands: Texel Island

New Caledonia: Kuto Bay, Île des Pins

New Zealand: Abel Tasman Park

Nicaragua: San Juan del Sur Beach

Nigeria: Coconut Beach

Norway: Hauklandsstranden

Oman: Shell Beach

Panama: Comarca Kuna Yala

Pakistan: Clifton Beach, Karachi

Papua New Guinea: Hisiu

Peru: Punta Sal, Mancora

Philippines: Bounty Beach, Malapascua; Palawan

Poland: Słowiński National Park

Portugal: Tavira Island, Algarve

Puerto Rico: Flamenco Beach, Culebra; Rincón

Romania: Costinesti

Russia: Curonian Spit

Saudi Arabia: Half Moon Bay

Senegal: Casamance Beach; Petite Côte

Seychelles: Anse Source d'Argent

Sierra Leone: Turner's Peninsula

Singapore: Tanjong Beach, Sentosa

Somalia: Berbera Beach

South Africa: Durban Beach; Long Beach, Cape Town

South Korea: Mallipo Beach

Spain: Islas Cíes, Galicia; El Palmar, Costa de la Luz

Sri Lanka: Unawatuna

St. Lucia: Reduit Beach

Sweden: Tofta, Gotland

Sudan: Port Sudan

Syria: Ras-al-Bassit

Taiwan: Kenting

Tasmania: Wineglass Bay

Trinidad and Tobago: Englishman's Bay

Tunisia: Djerba Island

Turks and Caicos: Grace Bay

Tahiti: Teahupo'o

Tanzania: Nungwi, Zanzibar

Thailand: Maya Bay, Koh Phi Phi; Patong Beach, Phuket Island

Tonga: Ha'apai

Turkey: Oludeniz Beach; Patara Beach

Ukraine: Arcadia Beach, Odessa

United Kingdom: Shell Beach, Isle of Purbeck, Dorset; Studland Bay, Dorset; Sinclair's Bay, Caithness, Scotland

United States: Alaska: North Beach, St. Paul, Pribilof Islands; California: Catalina Island; Florida: Sanibel Island; Hawaii: Honokalani Beach, Maui; North Carolina: Ocracoke

Uruguay: Punta del Este

Venezuela: Caya Sombrero Beach

Vietnam: Long Beach, Phu Quoc

Virgin Islands (British): Tortola

Virgin Islands (U.S.): Trunk Bay, St. John

Yemen: Al-Tair Island

3

OF WHAT ARE BEACHES MADE?
SEDIMENTS

Beaches occur in what seems to be an endless variety. Some beaches are so soft that your feet sink into the sand when you walk on them, while others are hard enough that you can drive and park your car on them without fear of getting bogged down. Beach colors range from black to yellow to brown to gray to white to green and even red. While sandy beaches change shape with every tide, some beaches are made of boulders so large that they hardly ever move. A beach may go on for miles and miles as an unbroken strip of sand (e.g., 90 Mile Beach, Australia) while others are only a few yards long in the midst of a rocky coastline. Some beaches are fronted by miles of tidal flats at low tide in locations where there is a large tidal range, such as along the Bay of St. Malo, France; Morecambe Bay, England; and the Bay of Fundy, Canada. In contrast, many of the world's beaches are just a few yards wide. Some that exist at low tide are nonexistent at high tide, especially in front of seawalls, esplanades, or rock cliffs.

Trying to make sense of this endless variety of beaches is the task of the coastal geologist or geomorphologist, and chapters 3, 4, and 5 outline some of the reasons for this variability. The differences between beaches lie in the combinations of many different factors. There are many clear and predictable differences between a beach in the Indian Ocean's Mauritius and another in Greenland; the most obvious difference is that it is pleasant to swim at one and not the other. In many instances, though, beaches separated by only a few hundred yards are vastly different. A passenger landing at Cardiff Airport in Wales will see a 250 yd (229 m)-wide fine sandy beach at the resort of Barry, right next to a 100 yd (91 m)-wide gray beach composed entirely of fist-size gray cobbles arranged

This oblique aerial view near Barry, Wales, illustrates how different two adjacent beaches can be. Sediment transported from the left is trapped against the rocky headland, while the darker sand to the right of the headland comes from a different source in that direction. The beaches differ in both material and grain size, from the light-colored beach of limestone cobbles to the brown, quartz sand beach.

Sand and gravel have been sorted by waves into two distinct sections on the beach at Streedagh, County Sligo, Ireland. Wind continues the sorting process, concentrating fine sand into dunes at the back of the beach.

into a series of descending steps like a staircase. The two are separated by only a 100 yd (91 m)-wide rocky headland. Some of the differences between beaches (such as the type of sand) are the result of local factors, while others depend on where a beach is located on the Earth. Beaches differ from one another because of climate (e.g., the tropics vs. the Arctic), geologic setting (e.g., a glaciated region vs. a volcanic island), and oceanographic setting (e.g., a beach facing the open ocean vs. one in a protected embayment). The differences that arise through variation in wave exposure can be seen by comparing the open-ocean beaches in the vicinity of Mar del Plata, Argentina, and the beaches of the nearby Rio de la Plata estuary, or the exposed east coast beaches of Barbados and those on the relatively sheltered west side of the island. Geologists sometimes differentiate leading-edge and trailing-edge coasts (see chapter 2). This categorization distinguishes beaches located on steep (usually rocky) coasts with a high degree of river influence (leading edge) from those on gently sloping coasts (usually with coastal plains), on which the continental shelf supplies sand and rivers play a much-reduced role in sediment supply.

At its simplest level, a beach is a sedimentary pile of loose grains that are sorted and shaped by waves, tides, and currents that are characteristic of a particular oceanographic setting. The sand, gravel, or boulders on a beach come from a variety of sources, the main ones being adjacent beaches, rivers, glaciers, the seabed, cliff erosion, and human activity. There are, of course, many different types of grains, and they come in an endless variety of shapes, colors, weights, mineralogical compositions, and sizes. Such diverse sediment properties create much of the variation among beaches, but not all of it, because the waves, tides, and currents that sort out these grains are also infinitely variable.

BEACH SEDIMENTS
GRAIN SIZE, SORTING, AND SHAPE

A beach is made up of millions of individual grains, each of which has its own shape, size, and composition. When combined, the size and shape of all the grains on a beach give the beach a distinctive *texture*. The texture of sediment on a beach is important for many reasons. For example, grain size determines the slope of a beach between the high- and low-tide lines, with larger grains forming steeper slopes. The response of a beach to spilled oil also will vary with beach sand grain size: Very fine sand permits very little penetration of the oil into the beach, while pebbly beaches have lots of spaces between grains, and these rapidly fill with oil. The communities of microscopic organisms that live between the grains of sand similarly vary with the size of the grains that make up their homes.

Grain size gives us our first impression of a beach. Sand is our common expectation, but beaches often are coarser than sand, ranging from pea- and pebble-size gravel to cobble and even boulder beaches. These coarser sediments are held onshore by wave energy, while fine sediments (muds) are held in suspension and ultimately are carried to deeper water. As noted earlier, mud beaches are rare, almost a contradiction to the term *beach*.

There are exceptions to the no-mud-on-the-beach rule. When a migrating shoreline, often on a barrier island, uncovers a compacted mud or peat layer that once existed in a marsh or river-bank deposit behind the island, the layer appears on the beach. These hardened mud layers may last several years on a beach and are usually taken away by storms eventually. During storms, waves may rip off chunks of the mud and roll them about to form *mud balls*. The same is true for peat layers, which are usually muddy, the result being *peat balls*. When a mud ball rolls around on a beach, coarser grains of sand and gravel are plastered onto the clay surface of the mud ball, giving it an armored outer surface that helps to preserve

This fine-sand beach on Xefina Island, Mozambique, has a gentle gradient. The building ruin is a World War II lookout bunker that was originally atop the rapidly retreating sandy bluff.

Upper left Mud layers exposed on the beach, Jekyll Island, Georgia, United States. The mud contains the roots of *Spartina* sp., indicating that it was deposited in a marsh. Marshes exist on the back side of barrier islands such as Jekyll, so in order for the mud layers to be exposed on the beach, the island had to migrate landward over the marsh. Such mud layers break off in chunks, which are then rounded into mud balls. Exposure of the mud layer is in part due to the seawall effect of reflecting wave energy and steepening the beach profile.

Upper right An armored mudball on the Sefton coast, England, that was eroded from Pleistocene deposits exposed on the beach. Small gravel pieces

became embedded in the mud as the ball rolled over the beach and are the armor that retards the rate at which the ball is reduced in size. The British fifty-pence coin is shown for scale.

Lower The darker-colored sediment on this North Carolina beach is mud, deposited in shallow depressions. Note the shrinkage cracks in the thin mud deposits (foreground) and the lighter-colored tongues of sand carried by the last round of higher waves that are burying the mud patches (background). This site is adjacent to Ocracoke Inlet, which accounts for the source of the mud.

the ball from further erosion (*armored mud ball*). Outcrops of clay-rich glacial tills, along shorelines in areas that were affected by continental glaciations, are a common source for clay-ball formation as the tills erode. The Gay Head Cliffs of Martha's Vineyard, Massachusetts, are a well-known source of variegated clay balls that become armored as they roll on the beach. *Tar balls* are analogous features—residual blobs of tar from oil spills or natural seeps are rolled about and become rounded and armored, like those on the shores of the Gulf of Mexico.

Tidal flats, the broad, flat platforms between the high- and low-tide lines, can be composed of sediments of varied sizes, but the largest deposits of mud at the shoreline usually are found in such flats or along the edges of lagoons. Examples of mud flats, with varying amounts of sand, include the Wash of eastern England and South America's Suriname shoreline north of the Amazon River. In contrast, the tidal flats of the North Sea's German Bight are very sandy. Mud also may be deposited directly on sandy beaches when water from adjacent muddy estuaries is ponded on the beach during high tide and the mud settles out.

Most of us have a good mental image of mud, sand, and gravel, but geologists use a common measurement scale to ensure that everyone knows precisely what is meant by the different terms that describe size. The Udden-Wentworth scale divides grains according to their diameter into several classes, including very fine, fine, medium, and coarse sand, all with precisely defined limits. Distinctions between sizes are made by sieving grains through known screen-mesh sizes, or by timing the fall of grains through a tall, water-filled settling tube, or simply by eyeball comparison of sands, with a card displaying known grain sizes used for reference. Material coarser than sand is divided into granules, pebbles, cobbles, and boulders. Pebble beaches are very common in high latitudes that have been affected by glaciers, which characteristically carry coarse material, and along mountainous coasts. Other examples of pebble beaches are composed of broken coral fragments, commonly found in the tropics.

On some shorelines, different-size grains accumulate in particular areas of the beach. In Ireland, for example, many low, sandy, intertidal beaches change to steep pebble beaches near the high-tide level because the pebbles roll across the sand, rather like bowling balls. On other beaches, grain size might vary laterally along the beach from fine to coarse sediment. On the 18 mi (29 km)-long Chesil Beach in England, the pebbles are much larger in the southeast (potato size cobbles) than in the northwest (pea size granules). Lateral variation like this may occur as longshore currents (discussed later in this chapter) pick up poorly sorted material from a bluff or river mouth and carry the finer-grained material for longer distances than the coarser sediment. Big grains also fragment during transport, leading to fining of grain size in the direction of transport. Some beaches show little sorting of sediments by grain size and simply consist of a poorly sorted mixture of sand and gravel throughout the beach.

To describe the range of grain sizes in a particular beach's sediment, geologists use a measure called *sorting*. This measure of uniformity or variability of the grain sizes

Upper left These basalt boulders making up the beach at Fingal Head, New South Wales, Australia, provide a good example of the boulder size category as well as material derived from a single source in terms of composition. These blocks were eroded from the adjacent headland.

Lower left This gravel beach on Oman's shore is dominated by pebble-size grains, but a few coarser cobbles are present on the back beach. During times when smaller waves hit the shore, the swash sorts out the sand fraction (the light patches on the lower beach). The beach sediments are derived from material eroded from the adjacent bluffs. Photo courtesy of Miles Hayes.

Upper right Good zonation of size sorting by waves is seen on this beach in Alaska's Pribilof Islands (St. George). Medium-size boulders are concentrated on the lower beach at the water's edge, the man is standing on a concentration of pebbles, and the ridge and upper beach are composed of cobbles and some small boulders. An erosional scarp, visible at the back of the beach, is the immediate source for some of the beach sediment, but the ultimate source is the uplands seen in the distant background.

Lower right A beach composed of coral rubble eroded from a coral reef offshore and transported onshore during storms at Beef Island, British Virgin Islands. The skeletal fragments of corals and associated organisms range in size from sand to cobbles.

Table 3.1   Udden-Wentworth Sediment Grain Size Scale

	Size Class	Size	Size Comparison
GRAVEL	Boulder	>256 mm	Basketball
	Cobble	64 to 256 mm	Potato to grapefruit
	Pebble	4 to 64 mm	Throwing size, skipping
	Granule	2 to 4 mm	Pea
SAND	Very coarse sand	1 to 2 mm	Peppercorns
	Coarse sand	0.5 to 1 mm	Coarse sugar
	Medium sand	0.25 to 0.5 mm	Granulated sugar
	Fine sand	0.125 to 0.25 mm	Caster/superfine sugar
	Very fine sand	0.0625 to 0.125 mm	Visible to the eye
MUD	Coarse silt	0.0310 to 0.0625 mm	Barely visible to the eye
	Medium to very fine silt	0.0039 to 0.0310 mm	Microscopic
	Clay	<0.0039 mm	Microscopic

NOTES: The term *clay* is used for both size and mineral composition. Clay minerals (composition group) are usually clay size.

ranges from very well-sorted sediment, in which nearly all the grains are the same size, to very poorly sorted sediment, which contains a wide range of grain sizes with no particular size being dominant. Sorting is affected both by the size range of the original sand and by the ability of waves, currents, and wind to sort the grains by size, shape, and density. Often the windblown sand at the rear of a beach is particularly well sorted because the wind selects out only fine sand sizes from those available on the beach. This well-sorted windblown sand will usually be finer than the coarser grains left behind on the beach as a *lag deposit*. Accumulations of very coarse shells and shell fragments are found on the surface of some beaches. These shells often originate as a lag deposit as the finer sand is removed, but the shells then form an "armor" on the surface that prevents additional removal of fine sand by the wind (see chapter 8).

Grains on a beach also vary according to *shape*. This shape variation in grains is particularly obvious on gravel beaches. Cobbles and pebbles, for example, are commonly divided into four main shape categories by measuring the lengths of their longest, shortest, and intermediate axes. This approach categorizes how closely pebbles resemble spheres, blades, discs, or rollers. Perhaps you have picked up one of these pebbles, flattened and disc-shaped, sometimes called a "worry stone," the ideal shape for skipping. The different ways in which different-shaped grains interact with waves result in their sorting across the beach. Discs and blades are easily lifted by waves because of their large flat surfaces, so they are often found high on the beach, where they

Upper left Beach steepness is controlled by grain size, with coarser material producing steeper beaches. This steep boulder and cobble beach at Annalong, Northern Ireland, consists mainly of rounded boulders (approximately 1 ft [0.3 m] in diameter) of Mourne granite piled high by storm waves. Author Andrew Cooper is standing just beyond the high-tide line, where the steepness of the beach increases significantly. The dark-stained boulders are the intertidal zone, which is not as steep because the boulders are deposited on a wave-cut platform in the underlying bedrock. The smaller cobbles and pebbles are very well rounded, and some are spherical. The dark line through the middle of the photograph is the wrack line, consisting mostly of seaweed.

Lower left An Antarctic beach dominated by boulder- to cobble-size material, but pebbles and sand-size material are present as well. These fragments of pink granite were deposited by icebergs and have not been extensively reworked and abraded as a result of wave activity, so the rocks are angular and generally not rounded. Contrast this picture to those of gravel beaches where the grains have been rounded by wave action. Photo courtesy of Norma Longo.

Upper right A gravel beach in Patagonia, Argentina, shows a variety of pebble types in terms both of shape and of composition, as reflected by their varied colors. Photo courtesy of Allen Archer.

Lower right An Alaskan beach made up of boulders, cobbles, and sand derived from the Malispina Glacier. These various sizes have not been well sorted by waves; however, the rounding of the edges of the coarser sizes suggest water transport and abrasion.

have been thrown by waves. Spheres and rollers are less likely to be lifted in the first place, but they also have the ability to roll back down the beach, so they are more common on the lower beach, close to the low-water mark.

Often, the shape of sand grains is initially inherited from the rock texture, the original mineral grain shapes, or the creatures (e.g., shells) from which the grains are derived. Mechanical abrasion then takes over as the shaping agent, rounding edges and corners of grains as they are transported by various stream and ocean currents and waves. Similarly, gravel-size material may inherit shapes from bedding, rock cleavage, or fracture patterns in the mother rock, which produce varied shapes. As they do with sand, the processes to which the rock fragments are subjected usually will impose the final shape. Slates, for example, tend to produce flattened rock fragments (shingles), while fractured granites are more likely to produce blocky cobble-size sediments. Abrasion by surf and waves will round the corners and edges of such granite cobbles to produce well-rounded and

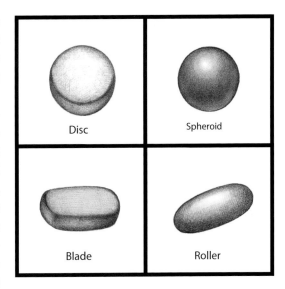

Simplified Zingg diagram to show some common grain shapes. Pick up an irregularly shaped pebble and imagine that internally it has three perpendicular axes—a long, short, and intermediate-length axis. The shape diagram is based on the relationships between these axes (e.g., in a sphere all three axes are of equal length). The shape categories are discs, blades, rollers, and spheres. Drawing by Charles Pilkey.

even spherical cannonball-size stones, well suited to be beach rollers. Shingles and other flat-shaped rocks (discs) often are deposited in an imbricate pattern, lying atop one another like the shingles on a roof. Pieces of animal skeletons that form beach grains also inherit shapes from the original part of the skeleton from which they are derived. For example, sea urchin spines and stick-shaped coral fragments become rollers, while some microscopic foraminifera tests (shells) create nearly spherical sand-size grains.

GRAIN COMPOSITION

Besides occurring in various sizes and shapes, the grains on a beach can be made of many different rocks and minerals. The most common beach sediment components are *terrigenous* (land derived from the erosion of other rocks), *volcanic* (directly derived from volcanic activity and volcanic rocks, as on many islands), and *biogenic* (composed of the shells and skeletons of dead marine organisms).

Although volcanic sediments may be land-derived by erosion, their distinction from terrigenous sediments is useful because in many island settings the beach sediment comes directly from volcanic activity or preexisting volcanic rocks. The black sands of Pacific Islands often are sand-size fragments of volcanic glass or fine-grained

The term *shingle* is applied to cobble beaches where the cobbles are flattened and rest on one another in a layered fashion, as do the cobbles on the upper part of this beach at Kettle Cove, Cape Elizabeth, Maine. The rock is a metamorphic phyllite, which is slaty in character, producing flattened slabs or disc-shaped cobbles when eroded from the outcrop (in the background). These cobbles were tossed onto the upper beach by storm waves.

basaltic rock, sometimes extruded below sea level, as off the big island of Hawaii, for example. Cobble-, pebble-, and sand-size fragments of explosive volcanic ash and rocks (pyroclastics) also are found on island beaches, as in the Caribbean, parts of the Mediterranean, the Pacific Islands, and some Alaskan shores. In the Mediterranean (Greece, for example), volcanic sediments may account for some red-colored beaches. Perhaps the most unusual volcanic beach sediment is *pumice*, a frothy glass with a fibrous texture that traps air in the rock. The result is a rock that floats, and pebble- to cobble-size pieces of pumice can be transported long distances to wash up on beaches far from the volcanic activity of origin.

The range of terrigenous grain types is as wide as the range of rock types on the Earth. By definition, a rock is an assemblage of mineral grains that are bound together as a result of crystallization (e.g., igneous and metamorphic rocks) or cementation (e.g., sedimentary rocks). A mineral is a natural compound with a characteristic chemical composition and crystal structure and characteristic physical properties; examples include the common beach minerals quartz and feldspar.

Most beach clasts comprise pieces of rock that have been rounded by transport to the beach or were derived from eroding sea cliffs or rock ledges on or near the beach. Most gravel-size particles on the beach are fragments of such preexisting rock. Occasionally even sand grains can be rock fragments rather than single mineral grains. Most sand grains, however, are composed of a single mineral.

Rock fragments provide clues as to the origin of the beach sediment. Some pebble and cobble beaches are made up of just one rock type, which is often an indication that the rock is locally derived (e.g., from erosion of a sea cliff or adjacent headland). Other coarse beaches have a wide variety of different rock types, suggesting mixed sources or perhaps a larger, more variable source area. On glaciated coasts, it is common to find pebbles that were carried long distances by glaciers or icebergs and mixed with many other rock types picked up along the way by the ice.

In Ireland, for example, pebbles of a unique rock, a granite containing a distinctive blue mineral known as riebeckite, can be found in beaches all along the east and north coasts. The source rock occurs only on the island of Ailsa Craig in western Scotland, so the occurrence of pebbles and cobbles of this unique riebeckite granite on the beaches of Ireland was important in helping early geologists conclude that glaciers transported the sediment from Scotland to Ireland. With the retreat of the ice sheets, these rocks, together with other, less distinctive rock fragments from all over northern Britain, were incorporated into the beaches.

As noted earlier, geologists categorize beach sediments into terrigenous, volcanic, and biogenic (carbonate) grains, depending on their origin. The composition of continental beaches is often expressed in terms of the ratio of the terrigenous and biogenic components to each other. On some islands such as those in the Caribbean Sea and the Pacific and Indian oceans, the ratio of volcanics to carbonates is useful in understanding the onshore and offshore sources of the beach sediment (e.g., volcanic activity versus reef erosion).

TERRIGENOUS SEDIMENTS

Terrigenous (land-derived) sand grains are the products of rocks that have broken down into individual mineral grains as a result of weathering and mechanical abrasion. Their mineral composition is diverse, but by far the most common type of sand grain in beaches is quartz, a light-colored to translucent mineral composed of silicon dioxide that occurs in igneous granitic rocks, metamorphic rocks such as gneisses and schists, and preexisting sedimentary rocks such as sandstones. Feldspars, another important light-mineral group, are the most abundant rock-forming minerals in the Earth's crust, but unlike quartz, feldspars are much more variable in composition (they are made up of aluminum silicates with varying amounts of potassium, sodium, and calcium). The feldspars are much less resistant to both chemical weathering and mechanical abrasion than quartz is, so the feldspars are preferentially removed. The result is that feldspar, in spite of its global abundance, is usually much less abundant than quartz in beach sands, although a small feldspar fraction is almost always present in terrigenous sands. Those few beaches that are rich in feldspar tend to be close to the source rock from which the feldspar is

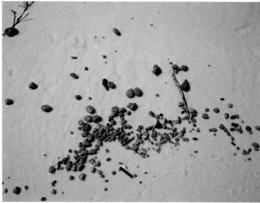

Upper left Microscopic view of a few grains of Hawaiian black volcanic sand. The well-rounded, lighter-colored grain is a sand-size rock fragment of basalt, and the black, glossy grains are volcanic glass, which is formed when lava comes in contact with seawater. The red bar scale is 0.04 in (1 mm).

Lower left This very coarse, poorly sorted beach sand from Redondo Beach, California, is a good example of beach sand in which many of the grains are fragments of preexisting rocks rather than grains of individual minerals. Most of these sand-size rock fragments are from a variety of rocks that were fine grained, relatively hard, and resistant to breakdown (e.g., volcanics, sedimentary chert), but individual mineral grains can be seen in a few of the rock fragments. Such diverse assemblages of sand grains are typical of beaches on the leading (mountainous) edges of continents and are characterized as composition-ally immature. Grains range from very well rounded to subangular, indicating their history of transport and abrasion, some having traveled a long distance to the beach while others were derived more locally. The polished appearance and roundness also reflect wave abrasion in the beach environment.

Upper right Pebble-size fragments of pumice on an Australian beach (Whitsunday Islands). Pumice is a natural glass produced by volcanic eruptions and has a low specific gravity because of its frothy texture, so much so that pieces of pumice will float on water, allowing currents to carry them far from their source. These pumice grains are somewhat dark in color, but pumice often ranges in color from a lighter gray to white.

Lower right Sand from a Batchawana Bay beach on Michigan's Lake Superior in North America reveals a history of glacial origins. The sand is compositionally immature, having been derived from the glaciers' eroding of crystalline igneous and metamorphic rocks (e.g., granites, basalts, and gneisses) of the Canadian Shield. Coarse sand–size rock fragments along with mineral grains of the dominant pink feldspar and white-to-clear quartz make up the sand. Note that the shape of the grains ranges from very angular to a few subrounded grains, and the lighter-colored quartz grains show some yellow-brown staining. This sand has not been subjected to the same degree of abrasion as has the sand in the Redondo Beach, California, example.

derived. For example, glacial deposits derived directly from so-called shield areas (exposures of very ancient rocks, such as the Canadian Shield, composed mainly of feldspar-rich granites and gneisses) produce beaches rich in feldspar. Some of the beaches of Lake Superior in North America are rich in feldspar. The average feldspar content in beaches of the southeastern United States, which are a long way (200 mi [more than 320 km] or more) from their source rocks in the Appalachian Mountains, is only 3 to 4 percent.

Any long time period and great travel distance will result in the quick breakdown of feldspar grains. When feldspars and other *silicate* minerals weather chemically, they produce clay minerals, which are also clay size—the source of mud. Of the common terrigenous rock-forming minerals, quartz is the most resistant to both chemical weathering and mechanical abrasion. So as other minerals weather to clays or are broken down into silt and clay-size particles, the sand-size fraction becomes richer in quartz. Thus, it is not surprising that quartz is the most common terrigenous beach material. The sparkling white beaches of the Pensacola, Florida, area have very high percentages of quartz sand.

Do not think, however, that all beaches are quartz rich. Most of the world's leading-edge beaches tend to be dark in color and have a wide range of grain compositions. Pacific Coast beaches of the Americas, from Alaska to Chile, and beaches of the western Pacific, eastern Indian Ocean, and parts of the Mediterranean that are near mountain ranges are often dark gray to dark green because the sand grains are rock fragments of preexisting fine-grained rocks such as metamorphic slates or volcanic rocks such as andesite and basalt.

Sometimes a beach's color is not the same as the inherent color of the beach's grains but results from the staining of the grains. The most common color of stained beaches is the familiar light brown or tan: The light-colored grains, such as quartz, feldspar, and even shell fragments, are stained by iron oxide, which gives these grains, and the overall beach, their yellow-brown color. In contrast, the sparkling white beaches near Pensacola, Florida, reflect the true color, or lack thereof, of the dominant quartz grains. Often the iron that stains the sand grains is furnished by the weathering of heavy minerals within the sand. However, such staining may have several different origins: It may have been formed from pore water in the beach; it may have been inherited from the grains' source (from oxidized soils, for example); or it may have originated while the sediment was on the inner shelf. Curiously, there are few studies of the character of such staining. One detailed study of stained quartz sands from beaches in southeastern Australia revealed (by examination using a scanning electron microscope) that the hydrated iron oxide (goethite) coatings were adhering not directly to the quartz grains, but to a submicroscopic clay coating. An abundance of microorganisms in the sand also can affect beach color (e.g., turning it shades of yellow and green). Perhaps the most extreme case of beach coloring originating from a source other than the sediment is in Brazil, where the water moving through the beach is so charged with iron in solution that it forms the red mineral hematite on exposure to air, forming areas of brilliant-red-colored beach where the water seeps out at the surface.

sand has not undergone severe transport abrasion, either before or since arriving on the beach.

Upper right This 100 percent quartz sand in which the grains are glassy, colorless, and unstained is an endpoint in beach sand composition and is said to be supermature because all other mineral constituents have been removed. Typically such sands are of fine to very fine grain size, they are well sorted, and the grains are rounded. Quartz sand beaches are most common on trailing-edge shores and are prized for their white purity.

Lower This heavy-mineral-rich, dark sand from Focene Beach, Italy, southwest of Rome, has a diversity of grain types, including glassy and white (reflective) quartz grains, possible light-colored feldspar, volcanic rock fragments (black), and a variety of heavy minerals (green and black), which reflect its source. The sand was derived from the Tiber River, which drains volcanic and terrigenous terrain. The fine sand is fairly well sorted (it has a uniform grain size), and the sample contains a few shell fragments as well. Some of these Tiber-source beaches were mined during World War II for the heavy minerals.

Upper left This fine-grained sand from a beach near Hiroshima, Japan, is an example of terrigenous sediment. Weathering of the island source rocks probably removed many of the unstable materials, so the resulting sand is dominated by quartz grains. Even the creamy white grains and brown-stained grains are probably quartz, although a few grains that may be feldspar, small rock fragments, and dark heavy minerals are present. Note that most of the grains are angular to subangular, suggesting that this

HEAVY MINERALS

Although light-colored minerals dominate, dark-colored minerals also occur on beaches. The most common are *heavy minerals*; the "heavy" definition derives from the fact that these minerals have considerably higher densities than both quartz and carbonate minerals. For example, quartz has a density of $2.7\,\text{g/cm}^3$ (water is 1), and heavy minerals range in density from 2.9 to $4.3\,\text{g/cm}^3$ for various types of garnet, to $3.5\,\text{g/cm}^3$ for diamonds, $4.6\,\text{g/cm}^3$ for zircon, $4.7\,\text{g/cm}^3$ for ilmenite, $5.1\,\text{g/cm}^3$ for magnetite, and $19.3\,\text{g/cm}^3$ for native gold. It is no wonder that early miners used panning techniques to separate gold from junk minerals on the basis of density differences.

Heavy-mineral sand grains (or "heavies") tend to be smaller than the associated light minerals in the same handful of sand. The size difference is because the smaller but denser heavy-mineral grains will behave the same hydraulically (e.g., sink at the same rate; require the same amount of energy to erode or be transported) as the equivalent coarser sand grains of common light minerals such as quartz or feldspar. The heavy minerals are the less abundant breakdown products of various rock types. Usually heavy minerals are scattered through the beach sand, typically making up 1 to 5 percent. Regional trends in their concen-

Groundwater discharging on this beach in Brazil has a high dissolved iron content, which forms red hematite on oxidation, giving the beach a vivid red color. Photo courtesy of Willard Moore.

trations often reflect the beach's proximity to the source of the sands and the processes that delivered the sands to the beach. For example, along the Atlantic Coast of North America the highest concentration of heavies is found in Canadian and New England beaches (e.g., Block Island, Rhode Island, beaches have more than 40 percent heavies, mainly ilmenite and garnet), where the sands came off the crystalline rocks of the Canadian Shield and were delivered into what is now the coastal zone by the continental glaciers. Farther to the south, the beach concentrations of heavies decline as the coastal plain widens and the distance to the ultimate sand source increases. Here the mode of delivery was by rivers rather than glaciers and the result was elimination of some of the minerals by weathering and abrasion. The decreasing trend continues to the south, and the beaches of the southeastern United States typically contain only between 1 and 2 percent heavy minerals. In contrast, there are beaches in Alaska and on some volcanic islands in which heavy minerals make up more than 90 percent of the grains.

On a beach comprising a mixture of light and heavy minerals, the heavy minerals can become concentrated by the sorting action of waves and currents into placer deposits, just like the placer concentration of gold in stream sediments. Look in the swash zone for a weak placering effect, streaks of dark heavies being sorted out from the rest of the sand. Similar thin placers may be found in the dunes at the back of the beach, and sometimes heavies will be sorted differentially in both water ripple marks and wind ripple marks. The thickest heavy-mineral deposits form during storms at the back of the beach, where storm waves break and then run up with great energy; the less energetic backwash can transport only light minerals, leaving behind the dark sand concentrates. In fact, on some beaches, black sands appear after every storm. Ditches dug by backhoes at the back of Gold Coast, Australia, beaches have revealed numerous black sand layers, each layer representing an individual cyclone.

Another process that concentrates heavy minerals is the formation of small anti-dunes by backwash (see chapter 6). The process can be an effective sorting agent and leaves a characteristic striped pattern on the beach as the backwash zone migrates across the beach on a falling tide.

Beach strollers sometimes mistake these black sands for oil pollution. Signs have been erected on the recreational beaches of Durban, South Africa, to reassure beachgo-ers that the black sand on their feet is a natural heavy mineral and not oil.

Some beach surfaces are composed almost exclusively of heavy minerals, most com-monly black sands that typically are dominated by magnetite (iron oxide) or ilmenite (titanium oxide). A simple test for magnetite is to pass a magnet over the (dry) black sand. If magnetite is present, the magnet will pick up sand grains. Other beaches are dominated by more colorful heavy-mineral sands, such as the red garnet sands of Lab-rador, Canada, or the green and red beaches of Lanzarote in the Canary Islands. Such colorful beach sand is formed when individual minerals are separated out from the rest of the heavy minerals by the sorting action of waves and wind. The red beaches of Lab-rador (Hutton Garnet Beaches), New England, and Bogue Banks, North Carolina, are the result of a concentration of garnet. Green beaches may be the result of the separa-tion of epidote (Outer Banks of North Carolina) or olivine (Pu'u Mahana Beach, Hawaii) from other minerals. The color may occur in the entire beach, as in the case of the oliv-ine beach in Hawaii, or in small thin patches, as in the case of the garnet on Bogue Banks beaches.

Careful study by mineralogists of such heavy-mineral sands shows that in addi-tion to a dominant mineral such as magnetite or garnet, there can be twenty to thirty additional mineral species, although many are present only in trace amounts. Like rock fragments, these heavy minerals in beaches tell us much about the source rocks from which the beach sand came. For example, minerals such as kyanite, stau-rolite, and sillimanite are from metamorphic rocks. In addition to the black, opaque minerals magnetite and ilmenite, other common heavy minerals include various garnet species, various amphiboles and pyroxenes, olivine, rutile, tourmaline, and epidote.

One unusual heavy mineral, which is light rather than dark in color, is the diamond. On the beaches of Namibia, diamonds behave in much the same way as the more com-mon dark heavy minerals. This characteristic helps geologists in their search for places where diamonds are likely to be concentrated by waves and currents; they look for the associated heavy-mineral deposits.

Another interesting mineral that occurs on beaches is mica. Mica characteristi-cally breaks into flat flakes that range in color from black (biotite) to purple (lepidolite) to golden to colorless (muscovite). It occurs in a variety of rock types, typically gran-

Upper Heavy-mineral layers (black) exposed in a trench on Whiterocks Beach, Portrush, Northern Ireland, separated by lighter-colored quartz-rich sand beds. The heavy mineral concentrations form as a result of storms, whereas the lighter sands accumulate during quieter conditions.

Center Wind rippling at the back of the beach on Bogue Banks, North Carolina, has caused some sorting by mineral composition, with a concentration of the wine-colored garnet on the steep ripple face, black magnetite on the lee face of the ripple, and light-colored quartz in the troughs. The 35 mm film canister is shown for scale.

Lower Heavy minerals are often concentrated in antidunes as they form in the backwash zone. On a falling tide this zone migrates across the beach, leaving behind a characteristic set of stripes, as seen on this South Carolina beach. Photo courtesy of Miles Hayes.

ites and coarse-grained metamorphic rocks. On beaches, mica grains are less abundant than quartz because coarser mica flakes are easily broken down into smaller particles and also because the flat mica grains are easily moved and often swept away from the beach into deeper water. Where mica does occur, it is either close to its source rock or adjacent to a river mouth that is contributing sediment to the shore. Periodic calm conditions favor the deposition of the mica grains, so, for example, many beaches of northern Queensland, Australia, in the shelter of the barrier reef, are covered in a fine layer of mica at low tide. The reflection of sunlight by the platy mineral surfaces can give these beaches a beautiful golden sheen. Similarly, beaches in the U.S. state of Georgia often sparkle from the reflective mica flakes, even though they make up less than 1 percent of the beach sand.

CARBONATE SEDIMENTS

The second common group of beach grain compositions is the carbonate minerals, which are discussed in greater detail in chapter 10. Carbonate grains are composed of the two common calcium carbonate minerals aragonite and calcite. Calcium carbonate is precipitated from seawater either in the form of sea creatures' shells and skeletons or, less commonly, as inorganic carbonate grains. These grains are transported by waves and currents and deposited on beaches, most typically in the tropics, although there are exceptions.

OTHER BEACH MATERIALS

Many other interesting natural and man-made materials form grains on beaches from time to time. Probably the most abundant human contributions to beaches are bricks. On beaches close to towns, it is common to find fragments of at least the occasional brick as well as other construction material such as tiles and cinder block. Often these have had the sharp edges knocked off as they rolled around in the surf, and some are completely rounded and barely recognizable as bricks except for their orange color. The

TERRIGENOUS BEACH MINERALS

A listing of the terrigenous minerals found in beach sands around the world. Some of the minerals on the list, such as amphiboles, garnets, and pyroxenes, are groups of minerals that include several distinct mineral varieties.

Light Minerals	Heavy Minerals (continued)
Quartz	Translucent to Transparent Minerals
Feldspars	Olivine
Orthoclase	Pyroxenes
Plagioclase	Amphiboles
Micas	Garnets
Muscovite	Epidote
Biotite	Apatite
Lepidolite	Staurolite
	Kyanite
Heavy Minerals	Sillimanite
Opaque Minerals	Tourmaline
Magnetite	Zircon
Ilmenite and leucoxene	Rutile
Pyrite	Monazite
Hematite and limonite	Topaz
Gold	Diamond

BEACH MINING

The BBC News headline read, "Jamaica Puzzled by Theft of Beach." In July 2008, sand thieves had removed five hundred truckloads of sand from its Coral Springs beach at the site of a planned resort. That this story made the international news was somewhat surprising, given that the destruction of beaches by both legal and illegal beach mining has been going on for more than a century in Jamaica and on other Caribbean islands. Most cases receive little attention beyond the local news. Perhaps the notoriety of the Jamaica story was due to the large amount of sand removed in a short time, and the economic impact of the beach loss at the construction site of a new resort. In fact, the authors themselves have witnessed illegal mining of beaches in the Caribbean region. While participating in an oceanographic cruise off the shore of Arecibo, Puerto Rico, one of this book's authors (Orrin Pilkey) observed just such an operation; it went on for all twenty-one days of the cruise. For eight hours a day, a backhoe continuously loaded dump trucks with sand, which was hauled away to an unknown location. A later visitation to the mining site revealed a nearby sign proclaiming that sand mining from beaches is illegal!

Beach mining is universal, and the largest legal operations have been associated with mineral extraction. Beach placer deposits of valuable minerals such as zircon, rutile, monazite, ilmenite, leucoxene, garnet, gold, and diamonds have been exploited in numerous countries, including Australia, South Africa, Canada, China, Norway, the United States, India, Brazil, Mozambique, Madagascar, and Senegal. Sometimes the mining operations restore the beaches after mineral extraction, but too often this is not the case. For example, beaches in southern Namibia and northwest South Africa have been degraded or destroyed by diamond-mining activities. In some cases, however, although during mineral mining the beach is significantly disturbed, the volume of sand actually removed is small. The "waste" sand can be returned to rebuild the beach. For example, the beaches of Nome, Alaska, have been mined for decades (one can still see sluice boxes on the beach, the property of individuals who hope to find gold-bearing sand that was missed in the gold rush). The actual volume of gold removed from the beach was minute compared to the volume of sand on the beach.

Sand is also used for the manufacture of glass and abrasives, but beach sand is most commonly used for construction aggregate. Legal and illegal beach mining is especially common in third world and developing countries, where rapid coastal development generates a high demand for aggregate used in concrete production or simply for fill. In a time of rising sea level, island nations are particularly hard hit by beach mining. In the Caribbean, virtually every island nation is affected. A 1995

Upper Local mining of a gravel beach in northern Sumatra is typical of third world mining of beaches for construction aggregate. Photo courtesy of Marianne O'Connor.

Lower This example of beach mining in Morocco is on a scale that significantly impacts the beach, both in terms of sediment loss that will lead to beach erosion and in terms of lost habitat. Photo © Lana Wong.

study of twenty Caribbean islands showed that beach mining was the second-most pressing problem for the islands, behind sewerage and solid waste disposal, and was closely associated with coastal erosion. Antigua has lost entire beaches to sand mining on its west coast. Anguilla's beaches have experienced narrowing due to illegal mining. Sand is so scarce on St. Lucia's coast that sand is now obtained by breaking up or crushing lava rock, a much less damaging source. Often beach mining is associated with low-income populations, but the systematic removal of dunes on Barbuda was carried out by a sand-mining company with which government officials were affiliated. When an injunction was issued to stop the sand mining, it was ignored. The minister of agriculture and mining-company officials were given short prison sentences, but the governor-general pardoned them. The mining of Barbuda beaches not only has resulted in beach erosion, but also has caused freshwater aquifers to become contaminated by salt water.

In rural Scotland and Ireland, where rocky glacial soils are common, farmers have for centuries taken sand from beaches and ploughed it into the soil to "lighten" it. The practice was probably relatively benign when done manually with a horse and cart, but the advent of tractors made the process much more efficient and damaging. Traditional "pebble dashing" of houses in Ireland and Scotland still sustains the (often illegal) practice of beach mining for small pebbles, which are scattered into the outer plasterwork of houses to create a distinctive wall covering.

Inland mining of creek and river sands that feed sand to the coastal beach system also impacts beaches, as is the case for beaches of the Placencia Peninsula, Belize, and beaches associated with the Rio de la Plata estuary in Uruguay. Montserrat banned beach mining in recognition of the problem, but locals continue the practice. In KwaZulu-Natal Province, South Africa, the impact of sand mining from rivers on beach erosion caused a heated debate in the local press in 2009. On Grenada (e.g., in Mt. Pleasant, Sabazan, Tibeau, and Grand Bay) sand mining continues to devastate beaches, and the issue pits citizens who attempt to protect beaches against politicians who say one thing and do another, truck drivers who ignore regulations, and builders who say they do not want to harm beaches, but continue to buy the sand. The Caribbean list goes on, with problems in Guadeloupe, Martinique, Puerto Rico, Trinidad and Tobago, St. Lucia, St. Vincent, and the Grenadines. The latter have imported sand from Guyana in an effort to reduce beach mining and meet the needs for construction aggregate, but this only transfers the problem to another locale.

The impact of sand mining on beaches and dunes is well documented in the Azores, where mining began in the 1960s. The mining was stopped in 1995 by legal enforcement, but the destroyed beach and dune systems have not recovered. In São Tomé and Principe, beach mining compromised the integrity of beaches and destroyed turtle-nesting habitat. Mainland Africa countries experience identical problems. Possibly the largest-scale beach mining in the world is in northern Morocco, where huge coastal dune fields are being removed as you read this passage. In a country with almost no lumber resources, sand for concrete is critical, but it need not be obtained from beaches. Lines of dump trucks wait their turn at the dune face; the trucks are quickly filled by an army of workers wielding shovels. Where steep slopes make truck access to the beach difficult, trains of donkeys string back and forth, each loaded to the maximum with sacks of sand. As many as 200 dump-truck loads of sand per day were said to be hauled from some mining sites. These operations were witnessed by authors Joe Kelley and Orrin Pilkey.

Beach mining in northern Morocco is on a grand scale; relays of large dump trucks carry away tons of sand daily. Note the protective barrier of sand left between the sea (to the left) and the mining operation. Ultimately that barrier will be breached, and the shoreline will move a significant distance landward. Although no buildings are threatened, the damage in terms of land and habitat loss is beyond recovery.

In 2007, increased coastal flooding in Liberia was blamed on coastal changes caused by beach mining, both for construction aggregate and for filling sandbags to build defensive structures against coastal erosion. From Cape Mount to Cape Palmas, coastal erosion is impacting housing development. In the adjacent country of Sierra Leone, the village industry of coastal Lakka was beach mining to feed building construction in Freeport; now Lakka faces an erosional threat from the sea because it removed its protective beach. Not far away, Benin is a good example of a small country with a coastal erosion problem greatly exacerbated by beach mining that was not controlled or regulated; it was driven both by high construction demand for aggregate and by economics. For a truckload of sand, beach miners could earn more than the country's average monthly salary, and beach communities collected a small fee for each truckload of sand mined, an income for the community. In 2008, the country was attempting to ban beach mining and shift mining to inland sand sources associated with rivers and lakes, but this created new problems.

On Africa's east coast, the rim of beach rock around the sandy barrier islands of Mozambique is the only material available for construction. The practice of mining this beach rock, however, is tantamount to digging out the very foundations of the islands. Elsewhere in the Indian Ocean, even the remote Andaman Islands experienced beach loss due to mining for construction in Port Blair. In the Andaman Islands and nearby Nicobar Islands, at least twenty-one beaches were reported to have been destroyed between 1981 and 2000, with the associated loss of turtle-nesting habitat and the protective function of beaches, as demonstrated by the 2004 tsunami. In 2008, it was reported that sand mining could erase some islands from the map in the Maldives; mining the coral sands from beaches and lagoons there

These piles of beach rock are being mined from the beach on Bazaruto Island, Mozambique. This beach rock helps stabilize the island, but it is the only rock available for use in construction on this otherwise completely sandy island. In effect, this is the contradiction of most beach-mining operations; the natural beach protection is being destroyed in order to obtain construction material to build buildings in the high-hazard coastal zone.

is a tradition, but it was accelerated after the 2004 tsunami, as aggregate was needed to rebuild and to fill sandbags for shore-hardening structures. In Sri Lanka, the impact of the 2004 tsunami was more destructive than it otherwise would have been because protective coral reefs from Akurela to Hikkaduwa had been mined for years to produce construction lime. Occasionally a local populace will make beach mining an issue, as in 2007, when people blocked a road in southwest India's Kerala Province because officials had failed to check unauthorized beach mining that had left the beach pockmarked with dangerous pits.

Pacific islands large and small have lost beaches and dunes to mining. Up until the 1970s, large volumes of sand were mined from Maui, Hawaii, beaches to provide not only aggregate but also lime for sugarcane processing. Sand was taken from Molokai to nourish beaches on Oahu, including Waikiki. Reportedly, replenishment beach sand for Waikiki Beach also has been obtained from Australia, Los Angeles, and from beaches on Maui, as well. Sand mining has been reported on various islands in Micronesia. Sand has been removed from beaches in American Samoa for golf courses, and an examination of sand traps on Hawaiian golf courses suggests similar sources there. Low-lying islands threatened by the sea-level rise cannot afford to mine away beaches, one of the last natural lines of defense against the coming storm and the very basis of their tourism economies, though small-scale mining by locals often isn't regarded as harmful. Sand is hauled away a horse-cart-load at a time, and the borrow holes fill with sand on the next high tide. It seems reasonable to conclude that no harm is done, but the net loss over time is real and damaging.

First world countries are not innocent of similar attitudes and mining impacts. Monterey Bay, California, and Jasper Beach, Maine, in the United States, the Kurnell Peninsula of Queensland, Australia, and Rodney, New Zealand, are just a few examples. Mining of North Stradbroke Island out of Brisbane, Australia, for rutile, zircon, ilmenite, and silica sands began in the 1960s and has destroyed frontal dunes. Although sand may be returned to the system after processing, the destroyed ancient Aboriginal middens and campsites as well as the ecosystems that were sheltered by the dunes are lost. In 2010, a mining company there proposed that an area of the island be added to a national park, but that company still retained mining leases on 45 percent of the island.

The attitude toward such environments in the 1950s and 1960s was that they were resources for the taking. For example, General Motors in the 1950s advertised new dump trucks, showing them being loaded with sand from North Carolina beaches. Although sand mining continues, there is a growing recognition that beaches and dunes are far too valuable in their natural state to be carried away in dump trucks or even in wheelbarrows and buckets.

A carbonate beach on the shore of Spencer Gulf, Australia, composed mainly of snail shells of various types. The Australian ten-cent coin is shown for scale. How many different genera of snails can you find in the photo?

flat, sandy beach at Crosby Point, in northeast England, has a very distinctive gravel beach at high tide made up of bricks eroded from an old dump site where debris from the World War II bombing of Liverpool was discarded. These bricks have been rounded and sorted by the waves to produce a cobble beach. In fact, waves can sort and form beaches from almost any material they encounter. At the fishing port of Portavogie, in Northern Ireland, decades of scallop fishing have seen the discarded shells mobilized by waves to form a beach almost entirely of scallop shells.

Beach glass, or sea glass (the "mermaid's tears" prized by beachcombers), is the broken-down and rounded remains of bottles and other broken glass that have been smoothed and sorted on beaches. These granule- to pebble-size glass fragments can be of various colors; the more colorful and unusual pieces are sought by beachcombers. (Beach-brick collectors are a much smaller group!) Another type of glass, reduced to sand size, has been produced from ground-up recycled bottles and used to nourish beaches in Florida and Hawaii; this approach has not met with public acceptance.

Rare human artifacts may also be found on beaches. Prehistoric stone tools are difficult to discern from other gravel- to cobble-size beach materials, but the lucky beachcomber may make such finds. Fragments of pottery and other ceramics that span human history make for exciting finds as well. Pottery fragments found in

A cobble beach at Crosby Point, England, composed of thousands of bricks, most of which have been rounded by wave action and deposited just like the cobbles on a natural cobble beach. The bricks are eroded from an old dump site dating from World War II; debris from the bombing of Liverpool was disposed of in this area.

some Tunisian beaches can be identified as Carthaginian, Roman, and every age since.

Often, waves can erode material that underlies the beach; when this happens, blocks of the eroded material are transported around the beach, rounded and shaped on the way. Cemented beach materials (*beach rock*), too, can be eroded and reincorporated into the beach.

Beaches can also be the final resting place of seaweed and sea grass that is ripped up during storms and carried to the beach. In western Ireland and Scotland, so much kelp is deposited on some beaches that they are completely covered. Usually this seaweed either decays or is washed out to sea. On recreational beaches the seaweed is often removed, but this can be harmful to the beach since the seaweed acts as a fertilizer for beach and dune plants, and many tiny animals live among the seaweed. Similarly, the

BEACH ROCK

Beach rock is a special kind of rock that forms when beach grains are cemented together by minerals precipitated from the water in a porous beach. The minerals grow as tiny crystalline needles on the surface of the sand grains, and eventually the crystals join, filling the pores as a cement that binds the grains together as rock. The process often involves rainwater, which is slightly acidic, dissolving calcium carbonate as it seeps down through the beach. When it reaches the seawater level in the beach, the carbonate is precipitated because the seawater is saturated with carbonate and cannot hold any more. Because this chemical reaction is faster under higher temperatures and because carbonate-rich beaches are more common in warm climates, beach rock is also most commonly found in the tropics.

Beach rock is usually found near the mid- to low tidemark as a solid rock layer on beaches exposed to a tidal amplitude of 3 ft or less (less than 1 m). The rock is made of grains that have exactly the same composition and shape as those in the adjacent loose beach sand. Some beach rock, we know, is very young. For example,

The raised area that looks like a roadway in the right of the photo is a beach-rock formation in Togo, West Africa. The rock is made of the same material as the beach except that the grains have been cemented into sandstone. In a sense, this formation forms a natural seawall, and like a seawall it reflects waves and lacks a beach on its seaward side. Photo courtesy of Miles Hayes.

The beach rock on this stretch of beach in the Whitsunday Islands, Queensland, Australia, virtually covers the beach and has changed the beach's recreational use.

Beach rock along the lowermost part of the beach near Lake St. Lucia on the Zululand Coastal Plain, South Africa, gives the appearance of a roadway, but it is a completely natural formation. One can imagine how such a feature might be mistaken for an ancient road or ruin. Note that there is a beach on the landward side of the beach rock.

These beach-rock slabs on the beach in Fujairah, United Arab Emirates, were probably deposited by very strong storm waves, or perhaps even a tsunami.

The large piece of beach rock in the right of the photo has been interpreted as a deposit from the 1755 tsunami that struck this shore at Cape Trafalgar, Spain.

along the north coast of Puerto Rico there is beach rock that contains beer bottle caps, nails, glass fragments, and even a crescent wrench, all firmly cemented in the sand. Beach rock usually forms slablike layers that are up to 1 ft (30 cm) thick and several feet (meters) wide and that dip in a seaward direction. These rock outcrops can extend for miles (several km) along a beach, as a kind of pavement. Indeed, the submerged beach rock on the banks off Bimini, Bahamas, which was left behind as sea level rose, has been mistaken for an ancient paved road from the mythical Atlantis.

Once formed, beach rock can undergo several different fates. The loose sand underneath the solid beach rock can be eroded out, undermining the rock layer, which then breaks into large slabs. These slabs can be moved by storms and thrown to the rear of the beach. At Cape Trafalgar, Spain, for example, slabs of beach rock were ripped up and deposited landward of the shoreline during the 1755 tsunami. Similarly, along the shores of the Gulf of Oman, which is bordered by the Gulf Emirates, an unknown event has thrown slabs of beach rock up from the beach. In another instance, a beach rock–eolianite deposit on the coast of Tunisia has been reported to contain evidence of a tsunami.

Beach rock may provide important habitat, as it is often colonized by plants and animals that otherwise don't occur on beaches. The very action of colonization can lead to the breakdown of the beach rock as organisms burrow and bore into it. Waves too can break pieces off the beach rock and transport these fragments as pebbles on the beach.

Undisturbed beach-rock layers often are found submerged offshore; or higher on the beach, above the existing low-tide line; or even landward of the beach. These layers mark fossil shoreline positions, so beach-rock lines are a good measure of sea-level changes. However, such nearshore ledges can also be a hazard to navigation. Such an underwater layer was responsible for the 2004 sinking of a marijuana-laden boat off Bazaruto Island, Mozambique.

A closely related rock type, eolianite, is similar to beach rock in origin, except the original sandy landform was a windblown dune. Again, natural cementation converted the dune sand to rock, in this instance most commonly in tropical areas. As sea level rose or fell, these fossil dunes formed resistant knobs, standing as small nearshore islands or small hills in back of the beach.

Beach rock can be the only

This eolianite outcrop at Black Rock, South Africa, forms a headland and is one of the few rock outcrops of cemented dune sands on this sandy coast. It provides a habitat for a small lizard that lives in the splash zone and is found nowhere else on the African continent. The lizard is confined to this single rock because it cannot compete with other lizards outside the harsh conditions of the splash zone.

The Black Rock outcrop of cemented dune sand in which the original dune bedding can be seen (above the boy's head).

solid rock on many tropical beaches. On the northeast coast of South Africa it forms seaward-protruding rocky headlands on an otherwise completely sandy coast. On Bazaruto Island, Mozambique, the authors have seen beach rock being mined for construction because it is the only solid rock on the island. When beach rock forms headlands and fixed points, the mobile parts of the beach can adjust to the new conditions imposed by the beach rock.

Upper left A beach in North Uist, Outer Hebrides, Scotland, covered in seaweed deposited during a recent storm.

Upper right The sea grass deposited in the wrack lines on this sand beach in the British Virgin Islands provides good evidence of the last series of high tides. The plant remains are an essential part of the food chain for beach and dune organisms.

Lower An incredible concentration of sea grass on a beach near Whyalla, South Australia. Such masses of decaying vegetation can generate gases that are unpleasant to beachgoers and can even be dangerous.

thin blades of sea grass can form thick accumulations on beaches. In some places, sea grass can be a major component of the beach. In the Spencer Gulf, in Australia near the town of Whyalla, is a beach composed almost entirely of sea grass more than several feet thick. Along parts of the Mediterranean coast, concentrations of marine vegetation are common; so much of the green plant material is in the surf zone that the breaking waves have an intriguing green color.

4

HOW BEACHES WORK: WAVES, CURRENTS, TIDES, AND WIND

THE MOST DYNAMIC PLACE ON EARTH

Our bias is that a beach never stays the same. The grains of sand as well as pebbles and cobbles on a beach are moved and sorted by a combination of waves, ocean currents, tidal currents, and winds. As the grains move, the shape of the beach changes, and so the next combination of waves, currents, tides, and winds work on a beach that is slightly to significantly different in comparison to the previous one.

Go to a beach during a storm and you will find a very different place compared to the same beach on a calm, sunny day (this is good advice for those imprudent enough to consider buying a house near a beach). Other, less dramatic changes take place on the beach when different types of waves and currents are at work. For example, the combination of a strong wind and strong current going in the same direction produces a very different effect from wind and current moving in opposite directions. On New Year's Day 1989, at Topsail Island, North Carolina, 40 mph (64 km/h) winds were blowing from the north exactly parallel to the shoreline, causing the water to move to the south like a mountain stream, complete with standing waves oriented perpendicular to surf-zone waves. The surf zone was a 150 yd (about 140 m)-wide swath of water discolored by its high sand content. Eight-foot (2.4 m) waves were breaking just offshore in the zone of rapid water movement, stirring up the bottom and adding to the current's sand load. During about eight hours of wind activity, the volume of sand moved in the surf zone probably equaled two or three years of normal sand transport

Plunging breakers like this curling crest form as waves move into shallow water. After the initial break, the wave re-forms and runs onto the beach. Note the curtain of spray rising off of the breaker, indicating a strong wind blowing seaward.

on this particular beach. Had the wind been blowing offshore, the sand transport would have been negligible.

WAVES

WAVE ENERGY

Waves are the single most important force determining the nature of a beach. Their ability to move sediment around and change the shape of beaches can be measured in terms of the amount of energy they carry. This energy transfer, in turn, depends very much on the height of the waves. During storms, when wave heights are at a maximum, much more sediment (and much bigger grains) can be moved than during normal wave conditions.

Waves are formed by wind blowing across the water surface. Wind transfers its energy to the ocean surface by friction as it blows, forming waves. The longer and harder the wind blows, the bigger the waves. The greater the distance over which the wind can blow to form waves (the *fetch*), the bigger the waves.

The waveform then moves that energy (not the water itself) through the sea surface toward the shoreline. Finally, the energy is expended on the beach when the waves reach the shore. The energy is spent or dissipated in three ways: the waves break, other types of waves and currents are formed, and sediments are moved.

Wave orbitals

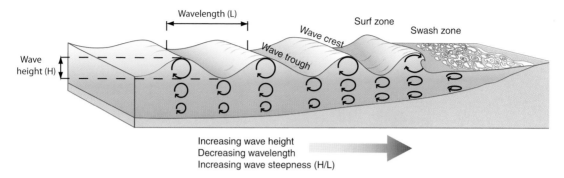

Diagram showing the changes in wave length and wave height as a wave approaches the shore. Water particle motion in deep water is circular (*orbitals*), with little forward motion of the water mass, but as the wave begins to drag bottom in shallow water, the orbitals become elliptical and the wave breaks in the surf zone, carrying the water mass forward to move beach sediment. The depth at which the wave first feels bottom is the wave base, usually about half of the wavelength. Drawing by Charles Pilkey.

It is an interesting fact that in deep water, beyond the surf zone, the water mass does not move forward with the wave; rather, the wave is simply energy passing through the water mass. As a wave moves through a particular spot, the individual water particles move in a circular orbital pattern, up and down but with only slight forward motion. The diameter of the orbitals of the water's motion decreases with depth until the wave is no longer disturbing the deeper water. In this way, the energy, not the water, is carried toward the shore. As the wave runs into shallow coastal water, the base of the wave intersects with the seafloor, begins to expend energy on the bottom sediments, and undergoes a change in wave orbital shape. This depth is known as the *wave base*.

Waves are described according to the *wave height* (the difference in elevation between the crest and the trough), the *wavelength* (the distance between successive crests), and the *wave period* (the time between two crests passing the same point measured in seconds). Wave buoys have been placed all around the world to measure wave character, and you can view records for many of these buoys on the National Data Buoy Center Web site at http://www.ndbc.noaa.gov. For example, such wave buoys off western Ireland record open ocean waves larger than 40 ft (more than 12 m) two or three times each year. In the wide expanses of the ocean, waves can get bigger than 50 ft (more than 15 m), but most beaches normally experience waves of 3 ft (1 m) or less.

The amount of energy in a surface wave on the open ocean is proportional to the wave height squared, so a small increase in wave height can mean a big increase in wave energy. A wave more than 13 ft (4 m) high, for example, has not four, but sixteen times more energy than a wave that is just over 3 ft (1 m) high.

These buildings are elevated on stilts at the landward edge of a very wide, flat, low-tide beach on the coast of Brazil. Note the yellow building and sign for reference points in comparison to this location in the high-tide photo.

The same location at high tide is a good example of how a small rise in sea level can move the shoreline position a considerable distance landward, a good analogy for understanding the impact of the longer-term sea-level rise. The low wave energy at this location does not put these houses at high risk; however, during infrequent storms considerable damage can take place. Photo set courtesy of Allen Archer.

Geologists often categorize beaches according to how much wave energy they normally receive (the Armstrong Price classification, discussed in chapter 2). Thus, we refer to beaches as low, moderate, or high energy. High-energy beaches are found on exposed oceanic coasts, such as the northwest coast of Ireland, the south coasts of Iceland and Australia, and the west coasts of Morocco, Taiwan, Chile, and the United States, while low-energy beaches are typical of sheltered seas or bays, such as those in southeast Florida (protected by the Bahama Banks), on the west coast of Turkey, on Spencer Gulf in Australia, and on the west coast of Madagascar. As noted in chapter 2, the width of an adjacent continental shelf also influences the wave energy reaching any particular open-ocean coast. The friction on waves that cross a wide shelf reduces the energy of the waves that finally break in the surf zone.

Storms can cause very dramatic changes on beaches. In the middle to high latitudes of both the Northern and Southern Hemispheres, low-pressure systems circle the globe, producing frequent storms. Beaches at these latitudes can expect to be affected by several storms every year. In the tropics, however, tropical storms, hurricanes, and typhoons occur seasonally but might make a direct hit on a particular beach only once every few decades. Right at the equator, few storms occur, which is why houses (on stilts) are built right on the beach, seaward of the high-tide line, as on the Gurupi Islands south of the mouth of the Amazon River in Brazil.

WAVE TYPES

Any visitor to the beach will occasionally see regularly spaced waves with long, continuous crests and similar wave heights approach the beach. These waves are called *swell*, and they are usually seen at the beach on a calm day. Swell, which is caused by distant storms, is most common on beaches that face a large expanse of ocean. Swell waves usually have large spacing between successive crests; the time between two crests passing the same point (wave period) is usually greater than 7 seconds and can be as much as 15 seconds.

On other days, the sea surface can look like the inside of a washing machine, with waves of many different sizes moving in several different directions in a chaotic pattern. These types of waves are referred to as *sea*. The difference between a sea and swells is really a matter of time and distance. When you find sea waves at the beach, it will usually be windy; a local wind has produced the waves on the spot. Sea doesn't have the beautiful continuous crests of swell; instead, the waves occur in discontinuous lines. These waves are usually also smaller (they have less wave height) than swell and have shorter wavelengths and wave periods (3 to 6 seconds).

At the point where the waves originally form by transfer of energy from wind to ocean, the waves are sea waves. As they move away from the point where they were formed, they begin to get sorted out according to size. Waves of similar size travel at similar speeds; each group is known as a *wave train*. The largest of the waves travel fastest, and they become the *swell*. Surfers know that along California beaches, swell may come from thousands of miles away, both from the Aleutians and from the Southern Ocean in the vicinity of New Zealand. Local weather forecasts from "surfers' stations" provide warnings of impending swells, sometimes days ahead of time, allowing surfers to make plans to catch these waves!

For large parts of the world, the waves are driven by persistent winds that come from a consistent direction, such as the trade winds. Examples include the waves on the outer coasts of the Windward Islands (e.g., St. Lucia, Martinique) and Barbados. In other areas, the wind direction is less predictable, and some areas have distinct seasonal variations in wind and wave direction. For example, the beaches on the East Coast of the United States are affected mostly by northerly winds in the winter and southerly winds in the summer.

The Hawaiian Islands are typical of island chains in the middle of an ocean basin. Depending in part on the time of the year, they receive swells from any direction, including the trade wind waves from the northeast, the Kona storms from the southwest, the North Pacific swell from the Aleutians, and the South Pacific swell from the Southern Ocean.

BREAKERS

Waves slow down as they move into shallow water because the waves start to interact with, or *feel*, the seabed. In this process of *shoaling*, the waves become higher and steeper and eventually become unstable and break. *Wave steepness* is defined as the ratio of wavelength to wave height. When the steepness gets to about 7, the wave becomes unstable and breaks. Large waves start to break quite far offshore, while smaller waves travel into shallower water before they break.

There are three types of breakers in the surf zone: spilling, plunging, and surging. The slope of the seabed and the size and steepness of the original waves coming in from deep water determine the breaker type. On gentle slopes, large waves (which are often low-steepness swell) usually break gently far from the shoreline as *spilling* breakers, the most common type of breaking wave. In this gentle type of breaking, the wave loses energy gradually as it continues to move onshore. Sometimes a single wave can form several lines of spilling breakers. This is the type of wave that often forms rip currents, described later in this chapter.

On steeper beaches with relatively steep waves, the crest of the breaker moves forward faster than its base, trapping air beneath it before it connects with the water again. This *plunging* breaker (sometimes called a dumping breaker) occasionally produces the "tube" of trapped air of which surfers are so fond. Sometimes these waves occur where there is a sudden rise in the seafloor, created, for example, by a large sandbar or the edge of a coral reef. Plunging breakers expend most of their energy all at once, unlike spilling breakers, which expend their energy while crossing wide, dissipative beaches.

On the steepest beaches, *surging* breakers form when waves rush up the beach in a smooth, sliding movement but no "curl" forms at the front of the waves. These waves can be dangerous to unwary swimmers, especially children, because of the strong currents associated with backwash. Surging breakers are most common on reflective beaches, where the incoming wave crashes into the steep, often gravel, slope.

After breaking, any remaining energy in the waves is used up in the *swash*, where the water rushes up the beach and either seeps into it or rushes back as *backwash*. It is this phenomenon that is responsible for determining the slope of beaches between the high- and low-tide lines (discussed further in chapter 5). As wave swash rolls up a fine-sand beach, very little of the water soaks into the beach. Thus, backwash can carry some of the sand in a seaward direction and flatten the beach. If the beach consists of potato-size cobbles, the swash moving up the beach in large part disappears in the crevices between the rocks. There is little or no backwash to carry sediment seaward to flatten the beach.

Usually the slope of the beach is just sufficient to cause all the incoming energy to be used up as the waves move onshore. During storms, of course, extra energy is in the waves and the beach might not be able to absorb all the incoming energy at the shoreline. When this happens, the extra energy is used up in different ways. Sometimes the waves erode dunes or cliffs at the rear of the beach, whereas at other times, waves pass right over the beach, carrying beach sediment as washover into the dunes or completely across a barrier into a lagoon or bay.

Giant waves up to 100 ft (30 m) high have been reported in the open ocean and are referred to as *rogue waves*. They form when waves coming from more than one direction combine to form a giant wave, which commonly disappears after a while as the component waves separate and continue on their original paths. On February 13, 2010, however, a rogue wave struck Mavericks Beach in Half Moon Bay, California, during a surfing contest. Not only were surfers upended, but onlookers were knocked down, and their cameras, cell phones, and backpacks were washed away. No deaths were reported, but a number of people suffered broken bones. Similarly, on July 3, 1992, an 18 ft (5.5 m)-high wall of water hit Daytona Beach, Florida, causing great damage to parked cars and slightly injuring seventy-five people. It is still not clear whether this was a rogue wave (weather conditions were calm at the time) or a rare North American Atlantic tsunami. If it was a tsunami, it probably formed as a result of a giant landslide on the continental slope 50 mi (80 km) offshore.

WAVE REFRACTION, DIFFRACTION, AND REFLECTION

One of the most important phenomena associated with waves is their capability to change their direction as they move into shallow water. This directional change happens because as a wave begins to feel the seabed, it slows down. If the wave is moving at an angle to the shoreline, then only part of the wave is feeling the seabed, while other parts are still in deep water. The part of the wave in shallow water moves more slowly than the part in deep water, and so the wave crest begins to bend. This process, in which the crest of the wave bends, or refracts, changes the angle between the waves and the shore.

Wave refraction is a complex phenomenon that depends primarily on the wavelength, the size and type of the wave, the direction from which the wave is coming, and the nature of the seafloor topography. Long waves (such as swell) feel the bottom in relatively deep water and so begin to bend far from the shore. It is quite common for swell waves to be completely refracted by the time they reach the shore, so that, although they may have arrived at the nearshore area at a steep angle to the beach, they break in a line of surf that is almost parallel to the beach. Shorter sea waves feel the seabed only in shallower water and may not be completely refracted at the shoreline, so it is more common for these waves to arrive at an angle of between 1 and 5 degrees to the beach.

Refraction around islands and across an irregular seabed can produce unexpected patterns as the resulting wave crests converge (concentrating their energy), diverge (dispersing their energy), and cross (creating turbulent conditions as the waves interact). In a classic 1947 study by Walter Munk from the Scripps Institution of Oceanography, the impact of continental shelf topography on waves on the beach was investigated. On the Southern California coast, Munk noted that submarine canyons cause waves to diverge on the beach, producing zones of low wave energy at the heads of canyons. A number of California fishing piers are located at the heads of submarine canyons, unknowingly taking advantage of the low wave energy produced by wave refraction in the canyon.

Waves can also experience a process known as diffraction, which involves a change in the direction of the waves as they pass through a narrow gap between two rocks, between jetties, or around a barrier in their path. This is the process that brings waves to the lee side of an object such as an offshore breakwater or other obstacle. It is the

Wave reflection is common off of rock cliffs and seawalls. Here at Portrush, Northern Ireland, the sea waves are approaching from right to left, and the spilling breakers are on the left of the wave crest. These waves are reflecting off of the seawall in the background, producing small waves that are spilling in the seaward direction (the wavelets visible in the middle of the photo, and the small wave in the far center background). These reflected waves are moving from left to right and collide with the incoming waves (note the colliding waves creating the large patch of white water in the center background). Sometimes a checkerboard pattern results, in which the wave-crest lines of the two wave sets cross. Photo courtesy of Norma Longo.

means by which the energy of incoming waves spreads laterally, perpendicular to the direction of the incoming waves.

Wave reflection is a process seen on rocky coasts, seawalled shorelines, and steep natural beaches. As the name implies, a portion of the energy of the wave is reflected and moves backward toward the ocean, through the incoming surf. Such waves may be partly responsible for the troughs in front of seawalls.

CURRENTS (LONGSHORE, ONSHORE, OFFSHORE)

Waves produce currents when they approach the shore. Probably the best known of these are longshore currents and rip currents. *Longshore currents*, sometimes called *littoral drift*, form when waves approach the shoreline at an angle and continue to push the water in the same direction as the wave breaks. Such currents are strongest in the

wave-breaking zone and are easily observed by anyone standing on the beach.
The greater the angle between the wave
crest and the shoreline (up to 30 degrees) and the larger the wave height,
the stronger the longshore current.
Winds blowing alongshore can also
cause or augment existing longshore
currents, as previously described for a
storm on Topsail Island, North Carolina. Tidal currents, produced purely by
the rise and fall of the tides, can also
create currents that flow along the shore.
Longshore currents are the principal
means by which sand is moved along
beaches in a process known as longshore
transport.

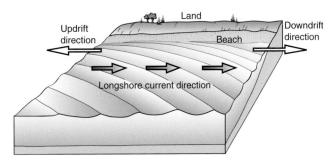

The longshore current is set up by waves approaching the shore at an
angle, in this case from the lower left in the diagram. The left end of
the wave crest line enters shallow water first, slowing down due to
friction with the bottom. Where the wave is in deeper water (to the
right), it continues without slowing down, so the wave crest line
appears to bend as it is refracted landward. The wave is piling water
up in the nearshore, and ultimately that water must flow laterally,
creating the longshore current. Such currents have velocities high
enough to move sediment set in motion by wave turbulence. Drawing
by Charles Pilkey.

The amount of sand moved on
beaches by longshore transport is highly variable from beach to beach. Such volumes
are difficult to measure or even estimate accurately, and the volumes of sand moved
vary greatly from year to year. For example, an estimated 10,000 cu. yd. (about 7,650
m³) are moved to the west on Bogue Banks, North Carolina, in a typical year. This number is so small and the accuracy of such estimates is so poor that it is considered to be
essentially zero annual transport of sand. More commonly, beach sand transport volumes on the order of 100,000 cu. yd. (about 76,500 m³) per year are estimated. The
net southward sand volume transport at Nags Head, North Carolina, is assumed to
be 500,000 cu. yd. (more than 382,000 m³) per year. The word *net* here is important
because there are many times when the wind, waves, and currents reverse and move
sand in the other direction. For the barrier island beaches of southeastern Iceland, the
estimate is a net of 4 million cu. yd. (well over 3 million m³) of sand moved to the east
every year. To understand the meaning of these numbers, remember that a good-size
dump truck carries 10 cu. yd. (7 to 8 m³).

Rip currents, always a hazard to swimmers, are narrow bands of water that flow
directly or obliquely offshore. They are caused when waves pile water in the surf zone,
and the water has to escape somewhere. Often rip currents occur through gaps in
an offshore bar. Under these circumstances, the currents will remain in one place for
hours and even days. The locations of long-term rip currents are occasionally marked
by indentations, holes, or depressions in the beach, a common occurrence in some Australian beaches. Sometimes rip currents are connected to longshore currents, dividing
the coast into a series of circulating cells with water moving alongshore until it joins a
rip current and then moves offshore. Other rip currents are caused by obstructions in

An example of a large rip current (center) on the coast of Tasmania, Australia, shows the seaward head of the rip, with suspended sediment in the water, and the zone where the current cuts across the surf zone, interrupting the lateral wave crest lines. Suspended sediment is also visible on the seaward side of the surf zone, suggesting that longshore transport is taking place. Onshore winds are building dunes on the landward sides of the pocket beaches. Photo courtesy of Andy Short.

the surf zone, such as rock outcrops or groins. Surfers often use the offshore flowing currents next to groins and jetties to float with relative ease out to where they can catch the next big wave at the swell-to-surf transition.

There are other types of off- and onshore currents that can also bring sand to and remove sand from beaches. When strong winds blow offshore from a beach, they gradually move water on a broad front (sometimes miles wide), away from the beach in a seaward direction. A depression is formed near the shoreline, crudely analogous to removing a cup of water from a water-filled bucket, and water moves ashore along the bottom to replace the water that has moved away from the beach. This landward-directed bottom current is referred to as *upwelling*, and if there is sand suspended by wave activity, this current may at times be strong enough to carry sand ashore, which is why offshore winds occasionally may cause a beach to widen.

The opposite may happen when strong winds blow in an onshore direction. The water piles up or mounds up against the shore, setting up a situation in which water must sink and flow offshore to maintain a uniform sea level. This mounding and onshore movement of water is called a *storm surge*, and the resulting seaward-moving bottom current, or *downwelling*, may take sand away from the beach. Hurricanes push water through and over barrier islands with the surge and onshore winds, but as the storm passes, the winds reverse to offshore and the amount of water moving seaward after the storm passes, referred to as ebb surge, can be very large and can have a very strong current. Sand can in this way be carried from the beach, many miles out to sea, even beyond the reach of the shallow nearshore waves that sweep bottom sand back onshore after storms. Such sand carried to these depths is lost to the beach system.

Each beach will have a different response to onshore and offshore winds, and observant longtime beach watchers (such as lifeguards and surfers) can often provide insight into how a particular beach works under various wind conditions.

Beautiful Lighthouse Beach on the coast of New South Wales, Australia, is fronted by a series of well-defined rip currents. Each break in the surf line of what appears to be calm water is a strong seaward-flowing rip current. The cuspate indentations in the beach that are lined up with the rip-current channels create a typical pattern. Note also that the two narrow rock headlands (top and bottom) are acting as natural groins, creating pocket beaches and blocking the longshore transport of beach sand. A mature dune field is at the back of the beach but has been reactivated (set in motion again due to loss of vegetative cover) in two areas by blowouts and associated parabolic dunes. Photo courtesy of Andy Short.

WAVE-CURRENT-SEDIMENT INTERACTION

Waves interact with the beach in a two-way relationship. The shape of the submerged part of the beach influences the way waves break and create currents, but the waves and their currents also cause the beach to change shape. They do this by sediment erosion, transport, and deposition.

More energy is needed to pick up a sand grain from the beach surface than is required to transport it once the grain is in the water column. Most grains are picked up at the point where the wave breaks and stirs up the bottom, a process easily observed by snorkelers. These grains are thrust into the longshore and cross-shore currents, which carry them for distances proportional to the current's strength. In the turbulent water of the storm surf zone, the grains may remain suspended for a long time, but under mild wave conditions, sand grains may move a few feet, then settle to the bottom, only to be kicked up again by the next breaking wave. The amount of sediment transport

Onshore bottom current

Offshore bottom current

Bottom currents in the nearshore zone can be generated by strong directional wind or wave patterns. Upwelling occurs when the wind pushes surface waters offshore, causing bottom waters to flow landward to replace the seaward-moving surface water. Similarly, when the wind or waves push surface water onshore, there must be a seaward return flow of bottom water to maintain the sea level. The result is downwelling. These processes are important to the ecosystem as well because upwelling brings nutrients back into the surface waters, and downwelling takes oxygen-rich surface waters into the deep. Drawing by Charles Pilkey.

and the direction in which it moves are, of course, highly variable, depending largely on the size and orientation of the waves.

TSUNAMIS

Some types of waves other than wind waves can affect beaches. Tsunamis (Japanese for "great harbor wave") are formed by sudden disturbances such as underwater earthquakes, volcanic eruptions, giant underwater landslides, or extraterrestrial bodies (asteroids) landing in the sea. The sudden release of a huge amount of energy and displacement of a large volume of water at the seafloor forms a tsunami wave, which travels rapidly (hundreds of miles per hour) across the ocean surface.

Tsunamis have very small wave heights in deep water, barely discernable by ships at sea, but wavelengths may be hundreds of miles. Just like normal wind-driven ocean waves, tsunamis slow down in shallow water and rise to great heights before breaking onshore. Japanese fishermen, returning from uneventful times at sea, have found debris and bodies in the ocean as they neared the shore, illustrating the difference between deepwater and shallow-water tsunami behavior. In the days before radio communication, the fishermen had no inkling that a tsunami wave had passed by them and struck their villages onshore.

The best-known example to the current generation is, of course, the Indian Ocean tsunami of December 2004, which had a wavelength of more than 124 mi (200 km) but was only slightly over 1.6 ft (50 cm) high in deep water. That wave traveled at 466 mph (750 km/h) across the ocean, and when it reached the shorelines around the Indian Ocean, it rose to heights of more than 50 ft (15 m). The volume of water in a tsunami wave is much larger than that in a wind wave of the same height, hence the immense damage. Videos of the event also indicate that much of the destruction resulted from surging currents that funneled up topographic lows and between obstructions once the wave broke against the mainland. There were strong currents set up by the backflow as well, both over the land and in the nearshore. The tsunami killed more than 230,000 people in eleven countries that bordered the Indian Ocean. Throughout the area, pre-storm human impacts, such as loss of coral reefs and destruction of mangrove forests to make way for shrimp farming, played a significant role in increasing the damaging effects of the giant wave on shorelines. While this book was being prepared, the Chilean earthquake of February 2010 struck and generated a deadly tsunami on the Chilean coast. Although the rest of the Pacific Rim beaches were little affected, the fact that people in Australia went down to the beach to see the tsunami arrive attests to the general lack of understanding about coastal processes. The Chilean earthquake demonstrated another natural process of beach modification, namely, the catastrophic uplift or downwarping of coasts.

Post-tsunami videos and photos, however, demonstrate an important fact: The beach is still present in most localities. Static structures are destroyed, but the dynamic beach flexes.

TIDES

Tides are formed by the gravitational pull of the moon and the sun on the rotating globe. For various reasons, the difference in water level between high tide and low tide at the coast is very important in shaping a beach. First, the higher the tidal range, the wider the beach is likely to be, simply because a greater distance of intertidal area is exposed. Second, if the tide is out, currents and waves do not reach the dry parts of the beach. At any point on a beach, the chances of waves moving sand depend on how often waves occur there, and these periods are reduced if the tidal range is great. Put simply, a large tidal range spreads out the energy of the waves over a wide surface area during a tidal cycle. Although wave action might be minimized on beaches in areas with large tidal ranges, beaches are still affected by strong, sediment-transporting tidal currents.

Some coasts have complex tides that rise and fall more, or less, frequently than the common, semidiurnal (twice-a-day) tides. On any coastline, tides also vary from day to day in a regular, predictable way as the Earth revolves around the sun and the moon revolves around the Earth. This variation, which fishermen and mariners are well aware of, is the change that occurs every seven days between the spring tide (largest tidal range) and the neap tide (smallest tidal range). Over the course of a year, the height of spring tides also varies, with the largest tides of the year occurring on the spring and autumn equinoxes (March and September): the equinoctial tides. There are many longer-term cycles in tides that are driven by planetary and lunar orbits and controlled mainly by the closeness of the sun and moon to the Earth. These include the 18.6-year lunar nodal cycle.

For coasts with a significant tidal range (more than 6.5 ft [2 m]), a storm striking a shore at high tide will cause a greater penetration of waves, water, and sand onto the land behind the beach than would the same storm striking the same beach at low tide. This difference certainly applies to hurricanes, which usually make landfall in a short time interval. However, nor'easters (extratropical storms with extensive storm fronts) usually last through one or more com-

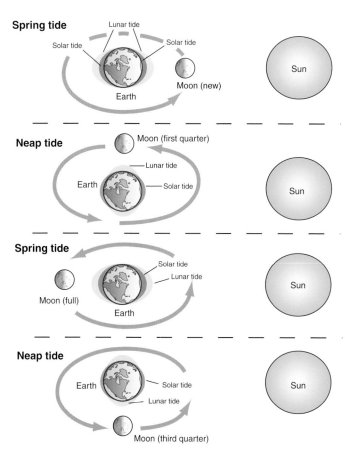

Spring and neap tides result from different alignments of the Earth-moon-sun system. Spring tide, which is the time of maximum tidal range between low and high tide, occurs when the Earth, moon, and sun are in a line so that the gravitational pull of the moon and sun are cumulative. When the sun-Earth-moon alignment is at right angles, the tidal range is at its minimum, the neap tide. Drawing by Charles Pilkey.

plete tidal cycles, so their impact is sure to be great for any coast that has a high tidal range. On the other hand, if the tidal range is small, like the 1 ft (30 cm) tidal amplitude of Tahiti, the timing of a storm makes little difference in its impact on beaches. A large (more than 26 ft [8 m]) swell on the KwaZulu-Natal coast of South Africa in March 2008 took place during the peak of the 18.6-year tidal cycle, on an equinoctial high tide. Large swells in the past had done relatively little damage, but the water level of this storm was raised on top of the very high tide, so the waves reached parts of

Although somewhat hidden in the shadows, this visible erosional impact and property destruction at Ballito on the KwaZulu-Natal coast of South Africa was caused by exceptionally high tidal levels during a 1997 storm. The waves atop the high tide not only undermined this house but also completely demolished another house that was in front of it.

the coast behind the beach that were normally well out of reach. The result was that this storm's impact was devastating, with widespread coastal erosion along what had been perceived as a stable coastline.

The third role of tides on beaches is the production of tidal currents, as water flows into and out of harbors, embayments, estuaries, and tidal inlets. Usually wave-formed longshore currents move much more sand than tidal currents do. However, strong tidal currents are common through inlets between barrier islands and in restricted embayments with large tidal ranges (e.g., the Bay of Fundy, Canada). These currents move large volumes of sand and form tidal deltas and various types of sandbars and fields of sand waves. Such sand transport and sandbar formation may add to or subtract from the sediment supply to adjacent beaches.

In general, the height of tides depends on the width and slope of the continental shelf and the configuration of the coast (see chapter 2). The flatter and wider the shelf, the higher the tides will be. Beaches can be found under a whole range of possible tides, from low-tide conditions (less than 1 ft [30 cm]), for example, those of oceanic volcanic islands and enclosed seas, such as parts of the Mediterranean, Baltic, and Caribbean seas, right up to the highest tides on Earth, in the Bay of Fundy, where the water level can rise more than 42 ft (13 m) between low and high tide. Other extreme high tides (with approximate highest tides) can be found in the Bristol Channel in southwestern England (more than 39 ft [12 m]), the Gulf of St. Malo in northwestern France (37 ft [11.5 m]), the Strait of Magellan in southernmost South America (36 ft [11 m]), the Okhotsk Sea in Siberia (more than 25.5 ft [9 m]), and northeastern Australia (25.5 ft [9 m]). Tidal range also is a controlling factor in the formation of barrier islands.

Coastal geologist Miles Hayes classified barrier islands, in part, on tidal ranges. Microtidal ranges (0 to 6.5 ft [0 to 2 m]) favor long, narrow (hot dog–shaped) barrier islands, while mesotidal ranges (6.5 to 13 ft [2 to 4 m]) favor shorter, drumstick-shaped islands. Barrier islands generally do not occur in areas of large tidal range (macrotidal, more than 13 ft [4 m]) because the strong tidal currents rework the sand into sandbars parallel to the direction of current movement, such as in the Bay of Fundy.

TIDES AND BEACH GROUNDWATER

The rise and fall of the tides is obvious in the visible surface-water levels, but beaches are saturated with water at high tide; they drain of that water as the tide falls. This changing of the groundwater level in the beach forces water and air into and out of the beach. As the beach drains during an outgoing tide, the water within the sediment is replaced by air. The incoming tide pushes the air out as the next tide rises. Evidence of such water drainage and escaping air is provided by the variety of small bedforms and sedimentary structures that result (see chapter 7). This effect also allows seawater and freshwater that enters the landward part of the beach from rain and runoff to interact and to produce yet another microhabitat for organisms.

OTHER WATER-LEVEL CHANGES
SURGES

Apart from waves and tides, the water level at the shoreline can rise and fall in response to surges. A surge is the difference between the predicted elevation of the ocean, based on the stage of the tide, and the observed height of the water. Surges are associated mainly with storms that have low air pressure and strong winds. The pressure is lower beneath the central part of a storm (e.g., the eye of a hurricane), and the water surface rises as a result (water levels rise 1.13 ft for each inch of mercury fall or 1 cm for each millibar decrease in pressure). The surge elevation is often even more strongly affected by strong storm winds piling water up at the shoreline. Hurricanes with winds of 75 to 95 mph (120 to 150 km/h) will typically produce a storm surge of about 5 ft (1.5 m), while winds of 130 to 155 mph (200 to 250 km/h) can create a storm surge as high as 18 ft (5.5 m). During Category 5 hurricanes, surges on the U.S. East Coast have reached more than 20 ft (6 m). Hurricane Katrina's surge reached more than 32 ft (10 m) in Mississippi. Even during a Category 1 hurricane, typical surges are 6 to 7 ft (1 to 3 m). The exact height of the storm surge is a function of the onshore wind velocity, the duration of the wind, and the slope of the inner continental shelf. The largest storm surge recorded appears to be 42 ft (12.8 m), which occurred on the coast of Australia in 1899.

Surges are largest in bodies of water that are shallow and confined by surrounding landmasses. Like tidal amplitude and wave height, a potential surge is higher when the shelf is flatter and wider. Thus, the Gulf of Mexico is prone to much higher surges than is the Pacific Coast of the Americas. Negative surges can also happen when high pressure or offshore winds occur. This means that tidal water levels do not reach the full elevation expected from the gravitational pull of the moon and sun.

EL NIÑO

El Niño is one-half of the phenomenon known as ENSO, or the El Niño–Southern Oscillation. This meteorological condition occurs when atmospheric pressure increases in the western Pacific (the area around Australia) and declines in the eastern Pacific (the area around Tahiti). The resulting change in the trade winds brings warm water to northwestern South America, changing a zone of upwelling into one of downwelling. In Peru, for example, this water mass is associated with enhanced storminess and a higher-than-normal sea level in an otherwise arid region. Once the coastal upwelling ceases and nutrients are no longer brought from depth into surface waters, the local fisheries collapse.

The storms and high sea levels of an El Niño event can wreak havoc on the shoreline. Along the heavily developed West Coast of the United States, stormy El Niño years are the occasion of greater and more damaging coastal erosion. In the early 1990s, a beach in front of a small village on the west coast of Colombia retreated 3 ft (1 m) per day during spring tides due to higher sea levels, raised by the El Niño effect. Although El Niño events do not occur with any regularity, they happen at least once a decade, and therefore are normal, if extreme, coastal processes.

WIND

Anyone who has been on a beach during a gale is inevitably amazed, at least the first time, at how much sand is collected by his or her hair. At the same time, sand grains will have blasted their legs. The ability of wind to transport sand on the beach depends on the grain size of the sand and how wet it is. Even wet sand can be moved. Studies have shown that when wind blows across a wet beach surface, a layer of sand one or two grains thick is dried out, allowing sand to be readily moved.

The wind can be responsible for the movement of large volumes of sand on the beach. The most obvious manifestation of the work of the wind is sand dune formation at the rear of the beach (see chapter 8), but wind blows along the length of the beach and into the water as well. The proof is in the sand accumulated in the lee of bottles, logs, and shells and a variety of surface features showing wind effects and direction. In short, the wind is a major player in shaping beaches.

Upper left Storm surges raise sea level during hurricanes well above the elevations of barrier islands, resulting in extensive flooding and overwash. Barrier islands in particular depend on such processes for their origin and evolution, as washover of sediment brings sand to the interiors and back sides of the islands. Houses, however, do not fare well when they are impacted by storm surges, as shown in this photo of Dauphin Island, Alabama. Note that even after a hurricane, the beach is still present, though it may have moved landward. Reports of "beach erosion" do not mean the beach has disappeared, though some of the houses have.

Lower left The Pacific coast of Colombia is retreating for a variety of reasons; however, during the sea level rise that occurs during El Niño events, the rate of retreat acceler-ates. This old plantation house was originally built a considerable distance inland, but by 1990 the shoreline was at its front door. On each high spring tide during El Niño, the scarp at the back of the beach retreated landward 3 ft (1 m) so that within a short time after this photo, the building was lost. The retreating beach remained.

Upper right Storm surges on any coast allow waves to reach parts of the beach that are normally beyond the influence of waves. Here the waves atop the surge water level are cutting into the toe of the sand dunes, forming a wave-cut scarp. Although infrequent, this is one way in which the beach adjusts during a storm; the dunes serve the important purpose of supplying extra sand to the beach and absorbing the wave energy.

Lower right The wind is sweeping this beach at Ostend, Belgium, parallel to the shoreline. It is blowing toward the bottom of the photo, as is demonstrated by the sand accumulating downdrift of shells in the foreground. Variable wind directions move sand landward, parallel to the shore, and seaward. If this were a natural beach, a changing landscape of dunes would probably be situated landward. Instead, a wall of high-rise buildings creates a static scene.

SEA-LEVEL RISE

The fact that the level of the oceans is rising is indisputable, whether humans are the cause of global warming or not. Tide gauge measurements going back one hundred years and satellite measurements going back to the early 1990s show that sea level is rising at a global rate of a little more than 1 ft (30.5 cm) per century. Most observers think that the rate is accelerating and that by the year 2100, the sea level will have risen somewhere between 3 and 5 ft (1 and 1.5 m).

During the twentieth century, the rise was caused mainly by thermal expansion of ocean water due to heating, plus a smaller contribution of water from both melting mountain glaciers and the Greenland ice sheet. As global warming proceeds, that relationship is changing: In the twenty-first century, meltwater from the West Antarctic Ice Sheet is expected to be the largest contributor, followed in importance by Greenland meltwater, the thermal expansion effect, and mountain glacier melting. If all of the West Antarctic Ice Sheet were to melt, sea level would rise 16 ft (nearly 5 m); the Greenland glaciers contain 20 ft (more than 6 m) of potential sea-level rise. All told, the mountain glaciers of the world have about 1.5 ft (more than 45 cm) of potential sea-level rise in them.

The actual amount of sea-level rise along the world's shorelines varies from place to place. On many of the world's river deltas, including the Ganges, Niger, Nile, Mekong, and Mississippi, the land is sinking, so the relative sea-level rise is large, locally as much as 4 ft (more than 1 m) per century in parts of the Mississippi Delta. At high latitudes in parts of Canada, Siberia, and Scandinavia, sea level is dropping because the land is recovering or rebounding from the release of the weight of glaciers that have recently melted away. Earthquakes often cause the land to sink or rise overnight, causing an instantaneous change in sea level, sometimes extending over hundreds of miles of shoreline.

Sea-level rise will do more than cause flooding. Storm waves will penetrate further inland; infrastructure will suffer serious losses (e.g., storm-water drainage systems, waste treatment systems); and salinization of groundwater and soils will occur. All of these effects already are occurring in some parts of the world. Ultimately, the sea-level rise will require massive retreat from the world's beaches. Where there is no development, the beaches will not be impacted except to change their location in space. Where buildings, roads, and seawalls line the shore, the survival of beaches may depend on whether humans try to hold the shoreline in place. If future generations decide to fight nature, the beaches are doomed.

Beaches thrive in a hostile environment where human structures fail and cliffs collapse. To do this, beaches find a balance between their size, shape, and arrangement of sediments, on the one hand, and the actions of waves, sea level, currents, and wind, on the other. Their ability to change shape as these factors change is the secret of their success.

Geologists take various approaches in their efforts to understand beaches. One way is to make a *sediment budget* calculation. Like a financial budget, a sediment budget involves deposits, transfers, and withdrawals. A beach in which the sediment budget is balanced is said to be in equilibrium. If there is a negative sediment budget, the beach will erode; if the sediment budget is positive, the beach will advance and widen.

Many beaches exist in a dynamic equilibrium involving the following factors:

· *Energy*—waves, wave-formed currents, and wind
· *Sea level*—the water elevation at which the energy is delivered to the coast
· *Materials*—sediments, including amount, type, and grain size
· *Shape and location*—the three-dimensional shape of the beach and the surrounding solid geology

When one of the four parameters changes on a particular beach, the others change accordingly. For example, if a dam is constructed across a nearby river, the amount of sand coming down to a beach (materials) will be reduced. Loss of sand leads to erosion (shape and location).

One of the most famous examples of a dam cutting off the sediment supply to a coast is Egypt's Aswan Dam. The Aswan High Dam, completed in 1970, almost completely cut off the sand supply to the beaches of the Nile Delta. Coastal erosion has ensued ever since, and the loss of nutrients brought to the delta floodplain and adjacent coastal waters has negatively impacted both agriculture and the commercial fishery. Similarly, in Southern California, between Point Conception and San Diego, dams on local rivers have reduced the sand supply (materials) by 50 percent for a number of beaches.

We know that the impact of storms (energy) may depend on the preexisting beach state (shape). Consequently, storms that strike in early fall will often make larger changes in beach shape than the same storm striking in early spring.

The gravel beach at the mouth of the Elwha River along the Strait of Juan de Fuca between British Columbia, Canada, and Washington State provides another example of how the equilibrium works. The present beach on the river delta is gravel, but before a dam was built on the Elwha River a hundred years ago, the beach was sandy, with only a small gravel content. After the dam was built, the supply of sand to the beach (materials) was almost entirely halted, but the beach depended upon a constant recharge of sand to maintain its grain size. Waves (energy) gradually removed the sand from the beach, leaving a lesser volume of gravel to accumulate to maintain a gravel beach that was narrower and steep (shape and location).

One problem that arose from the change in grain size was the demise of a clam habitat, eliminating an important source of food and income for a local Native American tribe. The Elwha Dam is scheduled to be removed in 2011 to restore both the salmon run and a sand beach for clam habitat.

And so it goes. The beach is one of the most dynamic environments on the Earth's surface and is often associated with even more dynamic tidal inlets. The beach changes rapidly according to the whims of the waves, but it changes according to a set of rules that we now understand in a general way. How exciting it is to stand on a beach, to understand what is happening, and to guess where the future lies. For example, we could have guessed that the California and Nile Delta dams would have led to beach erosion, and any coastal scientist today would predict that the Elwha Dam might cause the loss of the beach clams at the river mouth. Now we are faced with a global sea-level rise, and there is no question that the beaches will be moving back at an ever-increasing rate in coming decades.

While we know the general rules of beach behavior, we also know that each beach behaves differently because of the infinite number of possible combinations of nature's processes that affect them. This complexity makes beaches all the more exciting!

5

THE FORM OF THE BEACH: CRAB'S-EYE AND BIRD'S-EYE VIEWS

Even the most casual observer can't help noticing that beaches come in many different shapes. Some are steep and have a smooth surface, while others slope gently seaward and have lots of ridges and troughs. Viewed from above, beaches have a variety of shapes, ranging from long ribbons of sand that run along the shore for miles, to gently curving beaches strung between rocky headlands. Beach shapes are often represented in terms of their cross-sectional silhouette, the *beach profile* (i.e., a profile view; the way the slope of the beach varies from land to sea), and their plan view (i.e., a map view; the shape of the beach as viewed from above). Put together, the profile and plan views give us a three-dimensional impression of the beach. In other words, profile and plan views taken at different times—beach snapshots—can be compared, providing a way to express the continuous changes that beaches experience (see chapter 4).

BEACH PROFILES: THE CRAB'S-EYE VIEW

The vertical changes in beach elevation along a line running from the dry part of the beach through the low-water mark and out into shallow water is known as the beach profile. The profile records the irregularities in the beach surface and its elevation at each point and reveals different features in the form of the beach. The profile is one of the most commonly measured attributes of a beach.

We often think of the beach as just the part that we can see and walk on between the low-tide level and the landward dune or bluff, but generally beaches extend far under

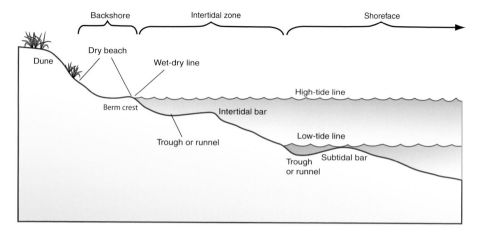

This cross section of an idealized beach shows the common beach zones and some of the major features at both high and low tide. Drawing by Charles Pilkey.

the sea. In fact, the beach profile extends from the point offshore where waves first begin to move sediment on a frequent basis, through the shallow water where waves break, through the intertidal zone, to the landward limit of the sand deposit reached by the largest waves. This final boundary includes the dunes at the rear of a beach, since these wind deposits sometimes contribute sand to the beach during storms and act as a long-term store of sand during fair-weather conditions. The beach profile can be steep or gentle, smooth or undulating. Many factors, such as tidal range, wave height, the occurrence of storms or long periods of quiet weather, and the supply of sand to the beach, influence the beach profile.

Although beach profiles vary a great deal, they have three distinctive sections that are found on almost all beaches. These are the *backshore* (the part of the beach above normal high tide), the *intertidal zone* (the part below the high-tide line that is exposed at low tide) and the *shoreface* (the part of the beach that is always submerged).

The shoreface is the underwater zone in which waves move sediment back and forth on the seabed. As waves cross the shoreface, they steepen and finally break as surf. The shoreface plays an important role in absorbing a portion of the waves' energy as they approach the coast. On some beaches the shoreface can be smooth and featureless. Indeed, it was once thought that all shorefaces had smooth, concave-upward profiles, but the upper portions of many shorefaces contain well-developed sandbars. Even though we cannot often see these offshore bars, their presence is revealed by the lines of waves breaking on the bar crests. Bars and waves work in tandem. The waves break because of the sandbars, and the sandbars are maintained by the breaking waves. Anyone who has swum offshore through the surf zone will be aware of the amount of turbulence on the bar crest. Often there are several sandbars, some of which will cause waves to break under all wave

Upper left The presence of an offshore bar is revealed by the line of breaking waves offshore of this southeast Alaska beach. Photo courtesy of Miles Hayes.

Upper right Multiple nearshore bars off the Massachusetts shore. Photo courtesy of Miles Hayes.

Lower Time-lapse photography is used to study the location and shape of sandbars. Here the time-lapse photograph of the waves breaking over a continuous sandbar paralleling the beach has smoothed the wave pattern into the creamy white zone that marks the offshore bar along Australia's Gold Coast. Photo courtesy of Gerben Ruessink from his research with the Water Research Laboratory, University of New South Wales, Australia, funded by the Gold Coast City Council.

conditions, while others in deeper water may cause the waves to break only at low tide or only during a storm. The North Sea coast of Denmark has as many as six offshore sand-bars, and the beaches of the western shores of Chesapeake Bay, on the East Coast of the United States, have up to ten. Nearshore bars can extend for miles along a beach, or they can be discontinuous. Some of these sandbars move onshore, "welding" wedges of sand onto the beach. Other bars move offshore, carrying sand into deep water. By using video cameras mounted on high buildings to record breaking wave patterns, scientists have been able to track changes in bar position and shape as wave conditions change.

The *intertidal zone* of the beach, the zone between the high- and low-tide lines, can be anything from a few to hundreds of feet wide, depending on the tidal range and the slope of the beach. We often think of our local beaches as typical in this respect and find it unusual if the intertidal zone of a beach we visit on vacation is very different from what we are used to. Tourists from the tideless Baltic Sea states, when asked about their perceptions of beaches in Donegal, Ireland, listed the rise and fall of the tide as their main dislike. They were not accustomed to walking hundreds of yards to the water and then having to retreat from the incoming tide.

A sand beach almost 1 mi (1,500 m) wide on the south end of Texel Island, Netherlands. The wide beach is a result of a high sand supply and is part of the Frisian Islands, the famous Dutch barrier islands. The tremendous width of the beach might discourage some, but not the intrepid coastal science field trippers.

A narrow beach in the Laguna Madre of Mexico results from a low carbonate sand supply, a low tidal range, and a sheltered position in the lagoon, away from high wave energy. The senior author, Orrin Pilkey, is shown for scale.

The intertidal zone is the most variable part of the beach; at high tide it is affected by wave and current action, and at low tide every day it is exposed to the sun, wind, and rain. On low-energy beaches, at the base of the swash zone, the backwash stops abruptly and the beach drops steeply by a few inches to a more gently sloping part of the seabed. This abrupt change in the profile is called a beach step. It commonly consists of coarse materials, and its formation is related to the waves' plunge point. Waves break on the step and send swash up the beach. The step is an area of high turbulence and is often marked by lots of sand in suspension in the water. For the same reason, it is usually coarse sand and shell debris that are concentrated at the base of the step.

As the tide falls, exposing the intertidal beach, water stored between the sand grains also tends to drain toward the sea, producing characteristic rills on the beach surface. The intertidal beach is often densely populated with burrowing creatures that emerge at low tide to scavenge on detritus left on the beach. On many northern Queensland, Australia, beaches, the surface is transformed within a few hours of the tide falling, from a smooth surface to one that is covered with the burrows and feeding trails of thousands of crabs (see chapter 9).

Some intertidal beaches are smooth and gently sloping, but many have one or more sandbars, which are usually smaller than the sandbars on the upper shoreface. Bars and their associated troughs that form on the intertidal part of the beach are known as ridges and *runnels* and are obvious at low tide on beaches such as the famous World War II Omaha Beach, in France. Ridges and runnels are common in areas with high tidal ranges, and more than ten ridges can occur across the intertidal beach. When the tide is moving out and the beach begins to drain, water is trapped in the troughs (runnels). This trapped water flows along the beach, using the troughs as channels, until it reaches a low point, where it breaks through a ridge and continues its flow down the beach. The ridges are often dissected by lots of these small drainage channels. As the tide rises again, each bar acts as a barrier to wave action for a while. Individual bars may even have their own swash zones and dry beach on the crests until the rising tide allows the sea to flow over the bar.

Berms are found at the limit of wave upwash at the boundary between the backshore and the intertidal zone. This upwash limit point is where the beach slope changes direction; one side of the berm faces the sea and the other faces the land. Sometimes the berm is moved above the normal high-tide line by small storms. Meanwhile, new berms may form as a result of sandbars migrating ashore. As the more landward berms become isolated from the active beach, they become "fossilized," or inactive. These former berms are called *beach ridges*, and on advancing (accreting) coasts there can be dozens of them. On parts of the Zambezi Delta in Mozambique, for example, there are more than fifty lines of beach ridges. Gravel beaches also can have several berms, which develop during different storm-water levels.

Beach ridge is a term with multiple definitions among coastal scientists. Some view beach ridges as the fossilized berms just described. Others define beach ridges as lines of sand dunes, such as the foredune (described next), which are more accurately called dune ridges.

At the landward limit of the beach, there are often dunes composed of wind-blown sand (see chapter 8). These dunes are of various shapes and sizes depending on the amount and type of sand they contain and the vegetation type and density of ground cover. Sometimes only a single dune line, the *foredune,* is present, but often there are multiple dune lines. These multiple dune lines usually indicate that the coast is advancing seaward (accreting). If large amounts of sand are being blown landward from the beach, randomly oriented dunes may form, rather than distinct ridges.

Occasionally, after a storm, the beach visitor will find a step cut into the beach or even the dune. This minibluff, or *scarp,* is formed by waves eroding the beach (berm scarp). Scarps can range from a few inches in height on the sand beach to many feet high at the leading edge of a dune (*dune scarp*). Often scarps that are cut during a storm will be repaired when conditions return to normal and sand moves back onto the beach. Artificial (nourished) beaches are particularly susceptible to the formation of erosion scarps. In fact, scarps are almost a diagnostic feature of such beaches and a strong indication of rapid erosion.

PROFILE CHANGES

By measuring the beach profile at the same position over time, scientists record the many changes in beaches that take place daily, seasonally, annually, and at longer intervals. Probably the longest records of this type of survey exist for the Dutch coast, for which there are more than fifty years of profile records. In the United States, Professor Jon Boothroyd, at the University of Rhode Island, has generated thirty-five years of beach profile data for that state. Such records help geologists understand the patterns of

(Captions on facing page)

Upper left A beach step forming on Peter Island, British Virgin Islands. Beach steps form on low-energy beaches, at the wave plunge point, as seen here just at the base of the swash zone. At the step, the backwash stops abruptly and the beach drops steeply by several inches to a more gently sloping part of the seabed. Waves break on this step and send the swash up the beach. This step is an area of high turbulence and is often marked by lots of sand in suspension in the water. For the same reason it is usually composed of coarse sand and may be the only place where gravel is seen concentrated on some sandy beaches.

Upper right Foredune line accumulating landward of the beach in Doñana National Park, Spain.

Center left Intertidal ridges and runnels (sandbars and troughs) exposed at low tide on the Sefton Coast, England. This coast has a tidal range of about 33 ft (10 m), and the beach is 0.6 mi (1 km) wide at low tide. As the high tide falls, the beach drains through the parallel troughs until the current is strong enough to cut across low spots in the bars (e.g., left of center). The crest of the middle bar has dried sufficiently here for the wind to begin to move sand (lighter-colored sand waves). Photo courtesy of North West England and North Wales Coastal Group.

Center right This dune scarp was cut by waves at Montrose, Scotland, and although erosional retreat is dominant, some natural dune regrowth can occur as the result of slumped clots of dune grass that reestablish their growth and aid in trapping windblown sand at the base of the scarp.

Lower left Multiple beach ridges of gravel in Dungeness, England, have accumulated during periods of abundant gravel supply when new gravel berms accreted to the shore, causing the coastline to advance seaward (left to right). The coarser berm crests (low ridges of gravel) have remained more or less unvegetated, whereas finer sediment in the troughs supports some vegetation.

Lower right Three berm crests can be distinguished on this gravel beach at Portmore, Donegal, Ireland, and size sorting has resulted in a sharp division between the lower sand beach and the storm berms of gravel.

beach behavior in response to various kinds of storms, wind and tide conditions, and changes in sediment supply. Another outstanding example is from Australia's Gold Coast, where Sam Smith, a coastal engineer, measured beach profiles and made other observations almost daily for twelve years for a particular beach that happened to be within a few miles of his home. His studies led to one of the world's best-understood beach systems in terms of how beaches respond to the seasonal dynamics of changing wave conditions, including the extreme conditions of storms and typhoons.

One of the interesting observations Smith made was in regard to processes in the swash zone. On many beaches and in many surf conditions, the swash exhibits a pattern of clear patches surrounded by thin lines of bubbles. Each of the clear patches is an orbital, as illustrated in the accompanying diagram. Smith referred to the pattern as "footballs." Although this phenomenon has not been investigated in any detail, it appears that the orbital action is a significant process by which sediment is picked up and moved within the swash zone.

Even without such long-term beach profile observations, geologists have noted the general rule that flatter profiles are associated with high wave conditions, and steeper profiles form under fair-weather conditions. As noted earlier, grain size is

Swash sloshing up against the face of a small wave-cut beach scarp. Note the foamy water being thrown over the lip of the scarp. This produces some interesting structures on the upper surface as well as some oddly shaped features on the face of the scarp (see chapter 6). Photo courtesy of Rob Greenberg.

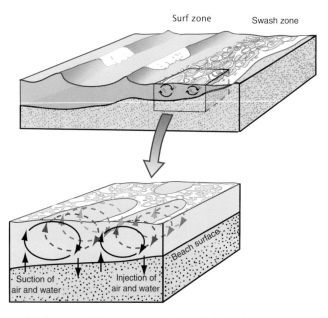

Surf zone   Swash zone

Beach surface

Suction of air and water   Injection of air and water

Model of Sam Smith's envisioned surf/swash "footballs," creating conditions that force a mixture of air and water into and out of the beach. Drawing by Charles Pilkey.

also important in determining the slope of the beach. Clearly, the beach's way of coping with different types of waves is to adjust its profile. As discussed in chapter 2, for a time these distinctive shapes were known as *winter* (storm) and *summer* (swell) *profiles*, but because of the way in which these different-shaped beaches coped with wave energy, they came to be called *dissipative* (all the energy was sucked out of the waves as they came onshore across a wide breaker zone with offshore bars) and *reflective* (all the energy was expended in one wave break plus the swash zone, and any remaining energy was reflected out to sea from the steep berm face).

BEACH PLANS: THE BIRD'S-EYE VIEW

An aerial view of the beach provides a different perspective about its form. This view reveals whether a beach is in a rocky embayment; on a spit, barrier island, tombolo, or cape; or in another environment. Within an individual beach, however, we can often see smaller features that form large patterns on the beach. Some of these patterns are on the high backshore parts of the beach, some are on the intertidal area, and some are underwater on the shoreface.

The turbulence of the breaking wave up into the swash zone (white water in midphoto to the left of the birds) has a surface pattern of football shapes that reflect three-dimensional cells mixing air, water, and sand in the upper layers of the beach. Trapped air and water produce various structures and patterns within and on the beach (see chapter 7).

BERMS AND CUSPS

Some beaches have a well-defined berm that runs for miles in an unbroken line along the beach. On others the berm crest has a scalloped appearance with a series of regularly spaced indentations, called cusps. *Beach cusps* appear as regular, rhythmic (evenly spaced) crescentric patterns formed by swash action. Each cusp is like a small embayment into the berm, bordered by "horns," or seaward-projecting points (small ridges), and the pattern is repeated along the beachfront for considerable distances, giving rise to the crenulated pattern at the beach edge. The scale of these features can be quite variable, ranging from the smallest cusps found on lake beaches (only inches to several feet in length), to swash cusps (25 to 80 ft [7 to 25 m] in length) and storm cusps as large as 400 ft (more than 120 m) in length.

Cusps are formed by nearshore circulation patterns and are particularly associated with storms. The details of their origin, however, are still in question. The protruding horns separate the troughs, or indented part of the cusps. The troughs can be more than a yard lower than the horns, and often the horns have coarser sand than is found in the troughs. At the center of each cusp, a small rip current forms as the water collects and then flows seaward. Cusps are most common on reflective beaches, and while they are found on both sand and gravel beaches, they tend to be larger on gravel beaches. It is common to see more than one set of cusps on the same beach. Cusps formed at a spring tide and then left higher on the beach on the falling tide may be less apparent, but if you walk down a beach and sense that you are traversing slightly up and down, you may be walking over a series of cusps.

The seasonal contrast in a beach is shown by this beach in southern Portugal. The 2007 summer beach shown here is sandy and has a gentler slope than the 2008 winter beach. Photo courtesy of Carlos Loureiro.

The same beach in the winter of 2008 is steeper and has a high gravel content as a result of the higher-energy winter storm waves. Photo courtesy of Carlos Loureiro.

A pocket beach in Kinnego Bay, Malin Head, Ireland, provides a good example of a well-developed set of cusps.

WASHOVER FANS

During storms, the level of the sea can reach an elevation high enough (a storm surge) that waves break over the beach and dunes and flow into the land behind. Such *overwash* commonly carries sand across the beach and deposits it inland as a *washover fan*. These are very common along narrow barrier islands and spits when there are gaps in the dune line. Washover fans can occur as solitary features or, more commonly, as multiple separate fans that line the back of a beach. When fans merge or coalesce, they create a washover apron.

The surfaces of washover fans can be covered in storm debris, which often provides the ingredients for the growth of new sand dunes when the storm has passed. By moving sand from the seaward to the landward side of a beach, the formation of washover fans is an integral part of the barrier island migration process. Along with dune migration, washover during storms is specifically how barrier islands migrate landward as sea level rises. Washover fans raise the elevation of the island, and if the storm is large enough, the fans extend all the way across the island and into the lagoon, widening the island. As the ocean-side beach erodes and the lagoon-side beach widens, the island migrates landward. Washover will become increasingly important as sea level rises.

OFFSHORE BARS

Offshore bars on the shoreface and ridges and runnels on the intertidal beach often occur in lines that run parallel to the coast. Sometimes, however, they take on more complex patterns. As previously noted, intertidal ridges on the beach are often broken by drainage channels that enable the water trapped in the runnels to drain from the beach on the falling tide. The water then flows into the next runnel seaward before finding another channel through the next ridge. This process produces a characteristic trellised pattern of channels running at right angles to each other across the beach. Although multiple ridges occur, typically a beach has a single prominent ridge.

Submerged offshore bars are difficult to observe directly, but their patterns can be seen from

An aerial view of washover fans on North Carolina's Outer Banks that formed during Hurricane Dennis (1999) shows how sand is carried across a barrier island to build up the back side of the island while the shoreline retreats. Combined with inlet formation and dune migration, washover is a natural mechanism for island migration, to keep pace with the sea-level rise. The beach is not destroyed in the process; it simply moves landward.

The view from a tall building on Australia's Gold Coast shows a line of breaking waves that is evidence of the presence of an offshore bar system; however, the pattern of that bar system cannot be determined by this instant image.

Time-lapse photography from the same location reveals the offshore bar pattern, which in this case is a set of crescent-shaped bars, concave relative to the shore (the light, creamy pattern). Photos courtesy of Gerben Ruessink from his research with the Water Research Laboratory, University of New South Wales, Australia, funded by the Gold Coast City Council.

the breaking wave patterns they produce. Using cameras mounted on tall buildings on the Gold Coast in Australia, and at Duck, North Carolina, for example, scientists are able to analyze the changing patterns of bar movement. Offshore bars can take on a whole range of rhythmic shapes. These include equally spaced small bars that run oblique to the shoreline at up to 45 degrees. Often rip currents flow between such bars and can be a significant hazard to unwary swimmers. Some beaches have dozens of these bars spaced anywhere from 33 to 330 ft (10 to 100 m) apart. Even when a near-shore bar runs continuously along the shore, it can be made up of a series of crescent-shaped sections. These crescentic bars are formed by the same types of regular currents that produce beach cusps. Headlands, small offshore islands, reefs, and shoals can cause wave refraction patterns that generate a variety of bars off adjacent beaches.

PART II

HOW TO READ A BEACH

In different climatic settings ranging from the tropics to the Arctic, natural beaches form as a result of the interaction of tides, waves, and currents with an array of coastal sediments of different compositions, grain sizes, and sorting. Add to this the effects of wind, of organisms that rework sandy beaches, and of the breathing of beaches as they fill and empty with water and air with the rise and fall of the tide. All beaches share similarities, but because every beach has been shaped by a unique combination of local processes in the last seconds, minutes, hours, days, and weeks, no beach is the same as it was moments before, or stays the same as it is right now. The beach as we see it is shaped by the present breeze, the last breaking wave, today's accumulated tracks of humans or of crabs and their burrows, the piles of seaweed and driftwood and sometimes plastic refuse, the last rise and fall of sea level on the tidal cycle, and the last storm. Being able to decipher the sequence of such events adds to our enjoyment of the beach.

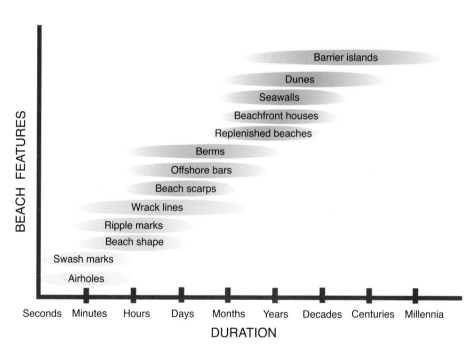

Comparison of the relative time spans for which various beach features tend to persist. Small features such as airholes and swash marks may last from only seconds to minutes, whereas large features such as dune fields and barrier islands may last from centuries to millennia under natural conditions. Drawing by Charles Pilkey.

Chapters 6 through 11 are a guide to reading the beach, focusing on features that range in size from the large (e.g., wrack lines and fields of ripple marks) to the smallest underfoot (e.g., foam traces and bubble marks). Internal features also are examined, from types of bedding produced by waves, to fabrics produced by air and water moving through the beach, to burrows, tracks, and traces of organisms.

6

THE BEACH SURFACE UP CLOSE:
IMPRINTS OF TIDES, CURRENTS,
AND WAVES

Reading a beach usually begins with looking at large landforms, such as the berm and offshore bars, and proceeds through examining intermediate-scale features, such as wrack lines and beach scarps, and then scrutinizing the small and even tiny features we see at our feet. Landforms and features seen in the topographic profile of a beach, such as large sand dunes or a narrow and steep backshore, tell the story of overall long-term conditions including sand supply, wave energy, and the impacts of engineering that have shaped the beach over many years or even decades.

Intermediate features, such as scarped or breached dunes, washover fans, and wrack lines at the back of the beach or into the dunes, are the evidence of the most recent big storm, which might have happened anytime from years to days ago. Multiple wrack lines or erosional scarps farther down the beach reflect more recent and smaller storms, or spring high-tide lines that may have formed only weeks or days ago; the lowest, fresh wrack line is usually from the last high tide.

Beach ridges and foredunes may reflect erosion-accretion cycles over a period of months or weeks, or longer time intervals in the case of gravel beaches. On sandy beaches, the berm crest nearest the sea often is the crest of the last sandbar to have migrated or welded onto the shore. On gravel beaches, the high berms recall the last storms of winter and remain out of reach of waves until the following winter.

The most interesting features, however, are often the smallest and most recently formed (in the last hours, minutes, or seconds), which collectively give the beach its surface character. This chapter focuses on these smaller features, the so-called sedimentary structures or bedforms on the beach surface. A *bedform* is the term for all small physical features on the surface of a beach or dune and can be thought of as any variation from a nearly flat surface.

NEARSHORE, BEACH, AND TIDAL-FLAT FEATURES

Nearshore water currents and waves expend their energy on the bottom sediments and shape the part of the beach that extends below the water surface. The sand often is reworked into various features that have a wavelike geometry (e.g., sandbars). Next to sandbars (see chapter 5), the largest of these features are referred to as *sand waves*, such as those formed in deep tidal channels. Although they are usually underwater, sand waves are visible in some aerial images of tidal inlets, and in areas with high tidal ranges they can be exposed at low tide on the margins of tidal channels. By definition, sand waves have crest-to-crest spacing greater than 20 ft (6 m). In shallower offshore areas, but where tidal and wave-generated currents are strong such as in the tidal channels, on the tidal deltas of inlets, and in the surf zone, *megaripples* can form. Megaripples are small, dunelike features with spacing between crests ranging from about 2 to 20 ft (0.6 to 6 m), and trough-to-crest heights of 4 to 20 in (10 to 50 cm). These bedforms cover extensive surfaces and migrate in the direction of the tidal or wave-generated currents. Megaripples are sometimes exposed at low tide on the surfaces of ebb-tidal deltas at inlets but are generally not seen on beaches. In cross section, megaripples are asymmetric, with a gentle up-current slope facing the direction from which the current comes, a crest, and a steep down-current face that slopes into a trough. Smaller ripples often cover their surfaces. Sand is carried up over the crest and avalanches down the steep face of the dune as it migrates. This deposition on the downdrift face produces inclined bedding within the megaripple, called *cross bedding*, in which the beds are inclined at up to 22 degrees, the angle of repose of medium sand in water (see chapter 11). The angle of repose is the steepest slope at which sediment will naturally come to rest when deposited on the wet surface.

RIPPLE MARKS

The most common wave- and current-created features found on beaches and tidal flats are much smaller than sandbars, sand waves, and megaripples, but they still show the wave or ripple form from which the structure gets its name: *ripple mark*. Ripple marks form an undulating surface of alternating ridges and hollows, which give the surface a corrugated appearance. Their crest-to-crest spacing is measured in inches, and their trough-to-crest height is generally less than 2 in (5 cm). The ripples form as a result of turbulence from a wave or current of water (or air) that comes in

Upper left Ripple marks are a common type of bedform. Tidal-flat surfaces are commonly covered by extensive fields of ripple marks of different types, along with animal feeding trails and burrow openings.

Lower left Submerged sand waves and tidal sandbars in the Parker River Estuary, Massachusetts. The waveform is common in sediments and ranges in size from these very large sand waves down to very small ripple marks. Photo courtesy of Miles Hayes.

Upper right Megaripples exposed at low spring tide near Cacela in the Portuguese Algarve. The waveform has an amplitude of about 2 ft (60 cm). Smaller ripple marks cover the surface of the larger megaripples.

Lower right Current megaripples exposed in Kosi Bay, South Africa. The wave amplitude ranges from 6 to 12 in (15 to 30 cm). Surfaces of the larger ripples have been modified by tidal currents to produce a lineation pattern.

contact with a bed of sand. The turbulence creates unequal forces on individual sand grains, which then begin to move by sliding, rolling, or hopping (saltation). Not all of the grains move uniformly or at the same rate, and some grains begin to accumulate, forming microridges, which cause more turbulence, allowing the ripple to grow in size. Sand grains move up the back side (the stoss side) of the ripple to the crest, and then are either carried off of the crest by the current or slide down the steeper ripple face in the direction of the wave or current movement. As a result, not only does the sand move, but the ripple form also moves in the same direction. The best way to understand this process is to watch ripples form and migrate in their natural setting.

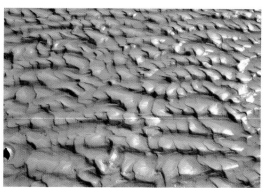

merge into a single ripple, typical patterns. The scale is 1ft (about 30 cm). Photo courtesy of Miles Hayes.

Center left Sinuous wave ripple marks showing the sharper crest of this ripple type, as well as parallelism of the ripples. Photo courtesy of Miles Hayes.

Lower left Current ripple marks vary in shape but are short crested and often arc shaped in plan view.

Upper right Very small standing wave in a drainage channel across a California beach. The underwater ripple pattern is the same as that on the surface of the water.

Upper left Nearshore ripple marks form as a result of wave action but usually show an asymmetry in their form, with a steeper face in the direction the waves were moving, in this case, from left to right. These asymmetric ripple marks show parallel crests, low amplitude, and a tendency for lighter debris to accumulate in the troughs. Note that a single ripple in the lower middle–right foreground branches into two ripples, while two adjacent ripples

Center right Surface waveforms in a creek mouth draining across a Puerto Rican beach indicate the bedforms developing on the sandy bottom. Note the white-water waves breaking upstream. These form over antidunes (sand is moving downstream toward the ocean, but the bedform is moving upstream). In the background, toward the ocean, there are smooth standing waves.

You can watch this happen sometimes in ankle-deep water in the swash zone or in deeper water using a mask and snorkel. Beach and tidal-flat ripples commonly develop in fine to medium sandy environments (grain sizes less than 0.02 in [0.6 mm] in diameter).

Two categories of ripple marks are found on beaches. *Wave ripples* are formed by the oscillatory (back-and-forth) current of waves and are characterized by parallel crests. A true *oscillation ripple mark*, formed only by the to-and-fro motion of the wave, is symmetric in cross section, with a sharp, straight crest and rounded trough. However, even when waves are a dominant factor in forming the ripples, there is usually a weak current, such as a tidal current, or a longshore current moving the sediment and causing the ripples to migrate. The result is that the wave ripples become asymmetric, with a steeper face developing on the downcurrent side of the ripple, but they still retain the parallel crests typical of their wave origin.

Most of the ripple marks found at low tide on beaches and tidal flats are of this type—mostly wave formed with the addition of the force of a weak-to-strong current. In a cross section cut with a shovel or trowel, the internal structure of the ripple shows cross laminae: individual layers or laminations of sand, often no more than a grain or two in thickness, that are inclined in the direction of the current's flow. The laminae in ripple marks are measured in small fractions of an inch, in contrast to the typical internal layers in the larger sand waves and megaripples, which are measured in several inches to feet (tens of centimeters to meters).

When wave energy is low and the water is clear, well-formed, parallel-crested wave ripple marks are often visible near the water's edge, just beyond the wave plunge point. The plunge point itself may be marked by coarser shell fragments or mixed sand and gravel, concentrated by the high-energy turbulence. Fields of ripples are best seen at low tide on exposed tidal flats. Although the ripple crests are more or less parallel, tracing individual ripples reveals that they commonly split (bifurcate) into two ripples, or two ripples will merge into one. This pattern results because a single ripple does not migrate at the same rate along its entire length. Somewhere along the crest line, the ripple detaches, migrates into the next ripple, and reattaches to give the appearance of either splitting or merging.

Current ripple marks result where a single-directional current (as opposed to an oscillating wave-formed current) is the dominant force transporting sand. The geometry of current ripples is significantly different from that of wave ripples. Most current ripples have short, discontinuous crests and are arcuate in form rather than parallel crested. When viewed from above, specific types of current ripples are named according to their shapes—linguoid (tongue shaped), cuspate (horn shaped), catenary and lunate (crescent shaped)—which are determined by features of the current's flow, such as its velocity and turbulence. These ripples usually occur in groups or sequences called *ripple trains*.

Current ripples typically form in channels where confined currents flow. On the shore, one of the best places to find current ripples and ripple trains, and to observe them forming and evolving, is in beach runnels (troughs) and runnel outlets. These troughs between bars and berm crests, when exposed at low tide, channel the drainage parallel to the beach, and strong currents can develop in these shallows. Current ripples form and migrate downstream as the current carries sand and finer sediment. In some cases, earlier-formed wave ripples will have produced sets of ripple crests parallel to the long axis of the trough, and then the later current will superpose current ripples atop the wave ripples (and vice versa on the rising tide).

Tidal creeks and creek or drainage channels coming off the land and crossing the beach also are natural observatories for seeing the relationships that link running water, sediment transport, and the resulting bedforms. In addition to current ripples, such channels may have larger bedforms during high-velocity flow. The presence of such bedforms is revealed by the character of the flowing water's surface. A smooth waveform known as a *standing wave* reflects a similarly shaped bedform on the floor of the stream. Where flow velocity changes abruptly, a hydraulic jump forms, like the white water seen at the base

Upper left Flat-topped ripple marks formed when the current reversed and moved sand toward the bottom of the photograph. The tops of the ripples were truncated, the sand was carried into adjacent ripple troughs, and in some cases the ripples were smoothed out in the direction of flow (left of the coin). The U.S. penny is shown for scale.

Upper right Flat-topped ripple field in which some of the ripples were breached and miniature channels carried sand into adjacent troughs to form small deltalike deposits. The U.S. quarter is shown for scale.

Center left Ladderback ripple marks produced by a strong set of ripples that moved from right to left; then a later set of smaller ripples was superimposed on top of the first set at a 90-degree angle, producing the appearance of rungs on a ladder. Note the fine shell debris in the ripple troughs. The U.S. penny (shown for scale) is resting against the steep face of the ripple.

Center right Interference ripple marks on the north coast of Spain show a right-angle orientation of weaker wave ripples atop the larger ripple set.

Lower left This set of interference ripples on a Brazilian beach produced a checkerboard pattern.

Lower right Flaser ripples form when mud fills in the troughs of the ripples. The ripples on this tidal flat at Yepoon, Australia, are sand, but the flooded troughs are filling with mud settling from the water. When buried and viewed in cross section, the sandy ripples will appear to be floating in a layer of mud.

of a dam. Where flow velocities are very high, breaking waves form and then appear to wash out, indicating the formation of bedforms referred to as *antidunes* on the stream floor. These underwater features can range in size from that of megaripples to that of ripple marks, with similar spacing, and are termed antidunes because the bedforms move upstream, even though the sediment is moving downstream. The smaller antidunes occur (very commonly) when they form under the very shallow film of water (backwash) rushing back down a beach following wave run-up (see later in this chapter).

(Captions on facing page)

As the tide or wind shifts during a tidal cycle and the current reverses direction, the crests of the wave ripples may be eroded off, producing *flat-topped ripple marks*, and here and there along the ripple crests, small patches of sand are carried into the troughs like miniature deltas or overwash fans. Second sets of ripples, related to different current conditions, can be superimposed over preexisting sets to produce checkered patterns (*interference ripples* or *cross-ripple marks*) or ladderlike patterns (*ladderback ripples*). Where mud accumulates, particularly in the ripple troughs, during times of quiet water between episodes of ripple formation, the resulting rippled sand layers will be interbedded with mud, producing *flaser ripples*. Other types of ripple marks form in the swash zone of the beach face, as described in the next section.

SWASH AND BACKWASH FEATURES

Watching wave run-up and the swash and backwash at the water's edge can be mesmerizing. A kaleidoscope of changing patterns and processes may be seen as the last bit of the wave's energy dissipates on the beach, either completely soaking into the beach or turning into backflow that runs back down the beach's face. More energy is spent in the run-up, which may have enough velocity and turbulence to roll shells and gravel-size particles, pick up sand, and transport sediment up the slope of the beach, than in the backwash. If backflow occurs, the water running back down the slope of the beach has less energy and moves finer material. Such winnowing sorts beach materials by size and, in the case of heavy minerals, by density (see chapter 3). However, if the beach is very steep and the waves are high, the energy of the backwash will be stronger than usual, and gravel-size materials and shells will be carried back downslope. Most pebble, cobble, and boulder beaches are so porous, however, that the swash usually soaks quickly into the beach itself and backwash occurs only during storms.

Gravel-size material is often sorted by shape as well as size. Strong currents also produce ripple marks in gravel. Flat pebbles and cobbles may be stacked in an imbricate pattern, like the shingles on a roof, and, as noted in chapter 3, when a steep beach is composed of gravel shingles and rounder materials, shape sorting occurs as the disk-shaped shingles are left behind on the top of the beach and the "rollers" return downslope. Sometimes the relative age of such sorting can be determined. At the rear of a gravel beach, it is common to find

Ripple marks form in gravel as well as sand, as shown here in Tierra del Fuego, South America.

pebbles that differ in color from those on the more seaward parts of the beach. Careful examination shows that these back-beach pebbles not only have weathered a little, but also are encrusted by vegetation, usually lichen. The extent of lichen encrustation can give an indication of how long the pebble has been inactive; the more lichen, the longer the inactivity. Similarly, on tropical gravel and boulder beaches composed of coral rubble, the most landward parts of the beach have been exposed the longest to the effects of slightly acidic rainwater. These coral fragments have been leached, as some of the coral skeleton has been dissolved, to produce very jagged edges on the coral fragments, which sometimes are stained black.

Near the water's edge, each swash event may deposit a single lamina, or layer, of sand grains, giving the beach its internal structure of parallel *laminae* (see chapter 11). These laminae are usually gently inclined seaward at the same slope of the beach face (less than 10 degrees). This swash run-up, with its loss of energy and ability to continue to carry sediment, combined with the bubbly foam at the edge of the swash deposits a con-

Shape sorting in beach gravels is common. Here the sliders (ranging from disc-shaped pebbles to small cobbles) are on the upper beach (to the woman's right), while the rollers (more spherical pebbles) are on the lower beach (to her left).

centration of material that marks the farthest advance of the swash. Such *swash marks* are one of the most universal of surface features found on beaches. These arcuate-shaped lines, which are stranded on the beach as the tide falls, record the last uprush of water on the beach face and usually show crosscutting relationships so that a sequence of swash events can be determined (earlier swash lines are truncated by the more recent swash line). These features are also common on the landward side of emergent sandbars

How long gravel particles have been on a beach sometimes can be determined by the growth of lichens on the surface of the stones. Here the "old" cobbles on the left are coated with lichen growth and appear darker in color, whereas the "new" deposit of wave-worked material (on the right) is fresher and brighter in appearance, lacking the organic surface coating. The pile of wire lobster traps marks the extent of the storm wrack line.

where a wave just barely makes it over the bar crest (or berm crest on the beach), again soaking in, leaving the material at the swash edge as a line on the back of the bar.

Swash marks are miniature wrack lines, enhanced by concentrations of heavy minerals, shells, and floating materials such as seaweed fragments, seeds, insect carcasses, and other fine particles of flotsam. The swash lines are often the focus of feeding activity by beach-dwelling creatures such as crabs and shorebirds.

Careful examination of the uppermost swash lines sometimes reveals another curiosity: evidence of *floating sand*. The swash line is usually a tiny ridge of sand, perhaps only the diameter of a sand grain or two higher than the surface of the beach. This microridge is due to a film of sand grains floating at the edge of the swash. Although sand is supposed to sink in water, surface tension can hold sand grains as a thin film on the water surface if there is no turbulence, and the swash mark is sometimes the deposit of this floating sand. Another location where floating sand can be found on beaches is at the edges of pools of still water in beach runnels at low tide; the sand typically appears as a scum on the water surface.

When the swash washes over the surface of the beach, it sometimes aligns or orients the individual grains so that their long axes are parallel to each other as well as to the line of the current. The beach surface appears smooth, but the parallel grains give rise to a streaked appearance, or *parting lineation*, and microridges and microgrooves terminate as lines more or less perpendicular to the swash marks. Because the swash curves in its run-up and because the grains may be reoriented by backwash, parting lineation sometimes shows a curved pattern, as if someone had taken a paintbrush or broom and made swooping brushstrokes over the surface of the sand.

Other seemingly mysterious surface markings result from any object moved over the sand surface by the swash where the object is in contact with the sand. These *drag marks* are as unique as the object being dragged along to scribe the mark in the sand, but common patterns are brushlike marks where small clumps of seaweed and algae are carried in the swash. If the object that made the mark is missing, the origin of the surface feature will remain a mystery, but sometimes the "tool" will be found at the end of the mark.

Backwash also modifies the beach surface, sometimes producing interesting patterns. One of these is the diamondlike or V pattern of *rhomboidal ripple marks* that form as backwash flows down a beach face of uniform sand. Steeper beach faces seem to favor their formation, and the Vs always point landward, or upslope. The one exception is where rhomboidal ripples form on the landward side of beach crests and sandbars that are overwashed by waves.

Similar V patterns, called *crescent marks* or *obstacle marks*, form as backwash flow scours around any larger, resistant object on the beach surface. The erosive scouring is due to the turbulence generated in the backflow around the obstacle, the same effect that gives waders that "sinking" feeling as a wave's backwash scours around their feet. Typical obstacles include individual shells, pebbles, and clumps of seaweed; the larger the object, the larger and deeper the crescent. Once an obstacle creates a disturbance in the flow, it is propagated laterally and downslope to generate a field of rhomboid ripple marks or similar pattern. For example, small patches of either crescentlike or rhomboidal ripplelike patterns often are generated where a group of burrowing organisms is just below the sand with their antennae or feelers protruding through the surface and disrupting the flow. Some burrowers build vertical tubes that are coherent enough to form obstacles when exposed, so the apex of some crescent marks is a hole.

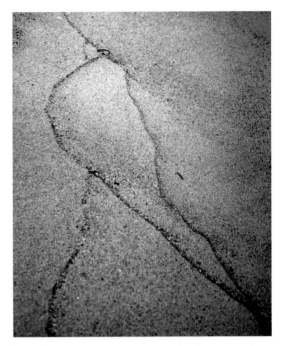

Swash marks form as each wave soaks into the beach, leaving the material it carried landward. These swash marks on a beach in Barbuda, West Indies, derive their color from the concentration of small pieces of pink coral at the swash margin. Each successive swash cuts the earlier swash marks, so a sequence of swash events can be determined. In this case, the mark in the lower left was the first event; it was followed by the middle-left arcuate mark, then the middle-right brownish mark, and finally the mark in the upper left. Each swash event left a layer of sand only one or two grains in thickness.

Any slight difference in the surface relief of the beach can create either linear scour patterns or linear depositional patterns without scour. Streaks of sand or microridges may extend in the direction of backflow from the lips of burrow holes or from other small objects that create a low-energy shadow in the downdrift direction.

Backwash down the face of a beach during a falling tide can produce the same high-velocity flow, called critical flow, that occurs in streams (described earlier in this chapter). The surface of the thin layer of backwash water will show riffles, or asymmetric water ripples, that reflect the small antidunes forming on the beach face. Although they are not as large as those seen in streams, the effect is the same. The ripple's steep face is

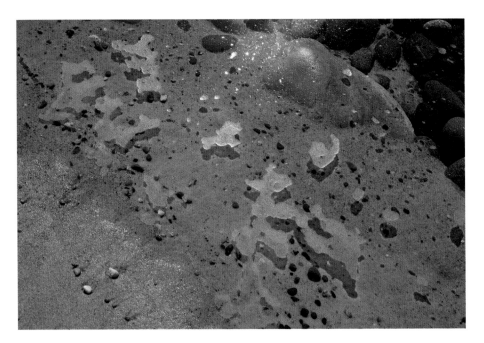

Floating sand appears as surface patches of sand on the quiet water in a beach trough on this Lake Superior (U.S.) beach. A thin layer of surface sand, no more than a sand grain in thickness, is picked up by rising water and remains floating due to surface tension. Where there is a gentle current, the sand will be carried in the direction of flow.

landward and is migrating up the beach slope, even though the sand is moving downslope (the opposite orientation of the current ripple marks). These bedforms are very short-lived and are almost always washed out by the continued swash and backwash activity, but the bases of these antidune ripples are often preserved, giving the beach a striped appearance. On beaches with high concentrations of heavy minerals, selective sorting of different-colored grains enhances the appearance of these features as light and dark sets of stripes on the beach. Where the higher part of the antidune dries out faster than the trough, a series of wet and dry stripes is also visible. At low tide, the surfaces of some high-energy intertidal beaches are completely covered in truncated antidunes.

The rising tide also produces some unique structures. Where waves first break over small beach scarps, return flow carries sand seaward back over the lip of the scarp. The small, saturated sand flows produce an array of intriguing patterns on the face of the scarp, as well as miniature deltas and fans at the toe of the scarp. These features are very short-lived: They are erased by the rising tide's waves as they attack the slope, or they collapse on drying if the scarp survives.

The high-tide waves may slop over such low beach scarps and produce irregular, thin patches of wet sand over the dry beach. There may be openings in the wet sand patch, which act as windows showing the dry sand in somewhat oval or circular patterns, with curled edges on the rims of the wet sand layer.

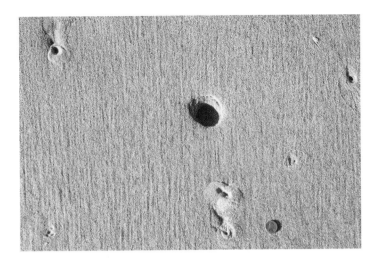

A strong swash lineation pattern is evident, oriented from top to bottom. The pattern is a result of the alignment of the long axes of the sand grains, although they are too small to see. Note also the truncated burrow hole and the sand extrusions around the burrow openings (see chapter 9). The U.S. penny is shown for scale.

A Puerto Rican beach in which the high content of dark heavy minerals helps to accentuate the swash and backwash patterns. Recurved backwash patterns in the black sand are prominent, seaward of the swash line, in the right part of the photo. Obstacle-mark patterns are also prominent as a result of backwash flow around the clumps of seaweed flotsam.

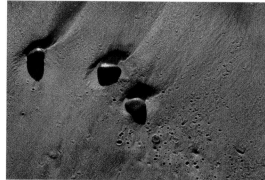

Upper right Close-up of obstacle marks resulting from pebbles on the beach surface. Backwash turbulence around the pebbles resulted in scour on the upstream side and then deposition in the direction of flow (toward the top of the photo). Note the nail holes due to air in the sand. The U.S. penny is shown for scale.

Lower right Rhomboidal ripple marks produce their characteristic diamond or V-shaped pattern on the surface of this Gold Coast, Australia, beach. The Australian two-cent coin is shown for scale.

Upper left Drag mark produced by a clump of seaweed dragged by a falling-tide current across the muddy sand of this Irish tidal flat. Drag-mark patterns vary with the different objects (tools) that leave their marks on the surface. Note the ripple marks, some of which are flat topped, the snail trails, and the mounds produced by burrowers. Footprints are shown for scale.

Lower left A large chevron obstacle mark around a pebble that is beautifully highlighted by the pink garnet sand fraction, selectively sorted by the backwash on this California beach. Note the faint rhomboidal pattern in the lower right. Photo courtesy of Miles Hayes.

Backwash riffles in the foreground indicate that very small antidunes are forming on the submerged surface of the beach. The backwash flow is seaward and is carrying sand seaward, but the ripples are moving landward as they form, hence the name antidune. These ripples wash out quickly, but differences in grain size and grain composition result in the striped pattern that indicates that antidunes were present.

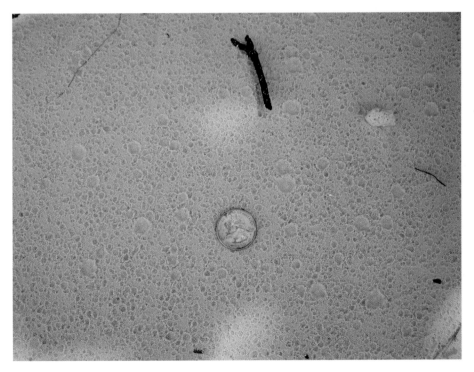

Raindrop impressions produced the pitted pattern seen over this area of a beach in the British Virgin Islands. Raindrop impressions tend to be fairly uniform in size, although they will show variation with raindrop size. The U.S. quarter is shown for scale.

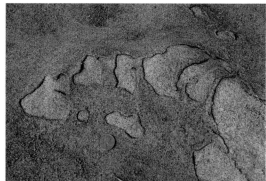

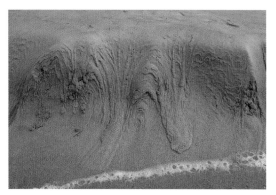

Upper left Airlie Beach, Australia, provides a good example of the characteristic striped-beach pattern resulting from the formation and then truncation of antidunes on the surface of the beach during a falling tide.

Lower left The face of a beach scarp where water from the last wave to break over the scarp has washed a slurry of sand down the face, producing interesting patterns of channels, fans, and deltas. The bubble-foam line is from the edge of the last wave. The U.S. penny is shown for scale.

Upper right The edge of the slosh from a wave throwing water onto the dry beach produces a scabby pattern from the thin, irregular layer of wet sand over dry sand. The lighter patches are dry sand showing through the windows of wet sand that curls along its edges as a result of surface tension. The same pattern often appears at the most landward edge of the swash line. The U.S. penny is shown for scale.

Lower right Drip marks forming from water dripping off boulders at the back of a beach in Santa Catarina, Brazil. The resulting pits are usually larger than raindrop impressions.

FOAM

Some features on the surface of the beach are so delicate and faint that they are difficult to see, but they are just as indicative of recent beach events as the more visible features. Such is the case with the marks produced by sea foam. *Sea foam* is formed during wave agitation of water containing organic compounds, which are added when waves churn up seafloor sediments or where tannin-rich river water enters the sea. Bubbles form and persist, creating the mass of foam, as in a bubble bath, and the foam is carried onto the beach by waves and wind. During big storms, foam can accumulate to thicknesses of several feet at the back of the beach, where it piles up against the dunes or other obstacles, including buildings. In August 2007, a foam layer formed along the Yamba, Australia, shore north of Sydney that was at least 6 to 10 ft (2 to 3 m) thick and led to the area's new name, "the Cappuccino Coast" (see www.dailymail.co.uk/news/article-478041/cappuc cino-coast-the-day-pacific-whipped-ocean-forth.html). A walk through such a mass of foam will leave a gray scum on one's clothing because inorganic clay particles are attached to the bubbles. Anywhere stormy seas are common, foam is common (in addition to the Australian coast, e.g., the U.S. and South American Atlantic

A patch of dissipating foam that moved from the base of the photo to its resting position. Note the faint parallel marks in the lower part of the photo. These are the inscribed trails or fine stripes created by the foam bubbles. The U.S. quarter is shown for scale.

and Pacific coasts, the coasts of the North Sea, the Japanese coast, and the South African coast). However, the average beachgoer does not visit the beach during storms and rarely sees foam at the height of its formation. Even during fair weather, though, small amounts of foam may be visible in the surf, and a strip of bubbles is usually seen at the edge of the swash as it runs up onto the shore. These bubble lines contribute to swash marks, as noted earlier. The bubbles of foam, as delicate as they are, leave not only traces, tracks, or trails as the swash or wind moves them over the surface of the beach, but also a pattern at the point where the bubbles dissipate. Because the bubbles disappear rapidly, they are rarely actually seen producing their delicate foam marks.

Where foam clumps are moved across the beach, either on a thin film of water or in direct contact with the beach, foam stripes are formed. These faint stripes parallel the wind direction. Occasionally the wind breaks off a large clump that is carried by hops, skips, and jumps across the beach, leaving an intermittent trace (foam track) each time it touches down. When the foam patch comes to rest and the bubbles burst, a final set of bubble impressions is left as a surface trace. These bubble pits are very shallow and of variable diameters. Although they are the most

Parallel bubble tracks on the surface of a Portstewart, Northern Ireland, beach. The British fifty-pence coin is shown for scale.

Foam skip marks (light patches in center) are the faintest of impressions left on the beach surface as the wind skips a ball of foam across the beach. The U.S. penny is shown for scale.

faint and delicate of features, once identified, foam structures tend to be recognized over much of the wet beach. When the foam dissipates, a film of dark mud and organic material can be left on the beach surface.

Another feature related to bubbles occurs at the wet-dry line during times when waves run well up onto the dry beach. Rather than the one-grain-thick lamination deposited by weak swash, a large wave swash will deposit a thicker layer of sand, though still less than an inch in thickness. If such a wet sand layer with bubbles is deposited over the dry beach, a scabby-looking surface develops when the bubbles burst, producing pits in which the dry sand can be seen below the wet sand. Surface tension causes curling of the wet sand around the rims of the bubble pits. These are similar to the features produced where waves slop over the crests of small beach scarps. Another common process that produces pitting is rain, but raindrop impressions differ from bubble pits in that they tend to be of uniform size and are distributed over much larger areas of the beach, as well as into any dunes that are present.

OTHER SURFACE FEATURES

Other common, but short-lived, features over the entire surface of the beach include *raindrop impressions*, which are microimpact structures, and shallow craters, often with a faint rim, the result of grains being thrown up by the drops' impact. Raindrop impressions can be confused with tiny depressions formed as a result of escaping air or water or collapsed air bubbles (see chapter 7), but these do not cover the entire beach or have the density of spacing seen in raindrop impressions. Similarly, *drip marks* from trees, ledges, or boulders produce pitted structures, but these are usually larger than raindrop impressions and are localized. There are also a variety of features produced by wind blowing over wet or dry beach sand (see chapter 8) and by animals leaving tracks, trails, and traces on the beach surface (see chapter 9).

7

ESCAPE FROM WITHIN: AIR AND WATER IN THE BEACH

If you are a frequent beach stroller, you have at various times seen small, startlingly perfect circles or rings a half inch or more in diameter on the beach. If you looked at the sand surface beyond the edge of your beach blanket, you may have noticed numerous small holes in the sand. On some parts of the beach, you might have noticed small mounds or blisters pushing up the sand, again spaced at regular intervals. And if you have walked a long way along a beach, you almost certainly have encountered patches of sand that were so soft that it was very difficult to continue your walk. If you were driving on the beach, you may have even gotten stuck in that patch of soft sand.

What do all of these observations have in common? All of these features are caused by the beach taking in and then exhaling air. A beach is alternately submerged, reworked by waves, saturated with water, and then, on the falling tide or after a storm, exposed to the air, allowing water to drain from the beach and be replaced by air. On the next cycle of submergence, water again saturates the beach, displacing the air. In the process, this bellows effect generates a variety of water and air escape structures. Even the most casual observer cannot miss the fact that the beach is "airy," as trains of bubbles stream out of tiny holes in the uppermost part of the swash on a rising tide.

The grains of sand that make up most beaches are imperfect spheres that, when packed together, retain open spaces between them. These voids, or pores, between the grains of sand on the beach are filled with water or air (along with microorganisms) depending on the level of the tide. There are two sources of water in beaches. The

most obvious is the ever-present ocean, with its crashing waves stirring up the sand grains. Less apparent is the flow of groundwater into the beach from the adjacent land.

On every shoreline except those in arid and Arctic climates, freshwater is constantly flowing into the sea through the beaches. The amount of flow depends upon the elevation of the lands immediately adjacent to the beach and on the local rainfall. Thus, a tropical beach on a rain forest–covered barrier island in northern Ecuador will have much more water flowing through it than will a beach in arid northern Chile.

Bubble trains in swash (lower right) appear as lines of small circles, the result of air escaping into the water as the swash passes over an air hole. The wave swash is just breaking over the berm crest, soaking into the beach. Note the line of concentrated foam at the edge of the swash; air and water are filling porous beach sand. To the left are clumps of foam where the previous swash soaked into the beach, displacing air in the pores.

This outward-flowing water could, in some instances, actually add to the rate of shoreline retreat. The theory is that the outward pressure on beach sand from the seaward-flowing water will make it easier for the waves to remove sand grains. This explanation is probably a minor factor compared to the effect of a high freshwater discharge saturating the beach and thus increasing the amount of backwash versus wave swash.

The amount of water that beach sand can contain is measured by the *porosity* of the sand, expressed as the volume of pore space between the grains. Typical beach sands have porosities between 20 and 50 percent, meaning that up to half of the volume of a beach is composed of air or water in the spaces between sand grains. The pores are interconnected, allowing water and air to flow within the sand, which is measured by *permeability,* or the rate at which water can flow through the sand. Pour a bucket of water onto the dry beach. The water disappears, soaking into the beach by moving through and filling the pores. Gravity pulls the water down until it reaches a level where it completely fills the pore spaces. The top of this zone of saturated pore spaces is the *groundwater table.*

Presupposing that a beach has perfect sorting of a single grain size, porosity is independent of grain size, but permeability is not. Fine beach sand allows much less water to flow than coarse beach sand or gravel because the pore spaces in fine sand are very small and there is greater friction between the pore water and the pore walls (the grains). A boulder beach has both high porosity and high permeability, which means that as a wave breaks on the boulders, much of the water is immediately absorbed into the interstices, or large spaces, between the rocks. The result is that the backwash from the wave is reduced because much of the wave water is lost in the boulders. On the other hand, a wave that breaks on a fine-sand beach loses little of its volume of water because the permeability of fine sand is low. Thus, almost all of the wave water returns seaward in the backwash. In addition, size sorting must be factored in when porosity and permeability are considered. Specifically, the smaller grains in poorly sorted sediments will fill in the pores between the larger grains, reducing porosity and contributing to the friction factor that increases resistance to fluid flow through the sediment (i.e., potentially reducing permeability). Fine- to very-fine-sand beaches tend to be well sorted, but coarser-grained beaches often show poor sorting. As discussed in chapter 3, these differences in permeability also explain why fine-sand beaches are flatter than coarse-sand or gravel beaches.

KNEE-DEEP IN SAND: AIRY BEACHES

Bubbly sand is produced by two natural mechanisms. The first process occurs when swash mixes air and sand, trapping air bubbles just below the beach surface. Trapping of foam may play a part as well. Hans Reineck noted that this occurs on beaches with steeply sloping upper intertidal zones that experience rapid sedimentation from swash, trapping air bubbles. This type of entrapment is also common on tideless lake beaches when air bubbles are concentrated in the berm crest, probably during small storms. When the sand dries, near-surface bubbles collapse, leaving shallow pits; also, one can stomp on the beach and cause a field of collapsed bubble pits well away from the footprint.

The second phenomenon that produces bubbly sand is more significant, in that the bubbly texture can be produced farther below the surface and can result in thicker layers of *soft sand*. It is also a much more widespread occurrence. The air entrapment in the beach results when air from the pores in the sand is forced upward through the beach on a rising tide. During low tide, the *water table* in the beach goes down and the pore spaces between the sand grains fill with air. As the tide rises, the water table in the beach also rises, and the surface sand becomes wet, so the air in the beach is trapped between layers of water-saturated sand. The air is forced upward, and some escapes through sand holes in the

Collapsed bubble holes just below the surface of the beach. The holes resulted from the mixing of air and sand in the swash zone, possibly associated with foam. Typically, these are concentrated at the berm crest, and stomping on or simply walking on the beach will generate their collapse and the resulting pitting. The U.S. penny is shown for scale.

Bubble sand seen in cross section gives the impression of Swiss cheese. This Brazilian beach shows not only the variation in the size and distribution of the holes produced by trapped air in the beach, but also some movement of that air (vertically elongated holes).

beach, but discrete air bubbles also form in the sand and are trapped by overlying layers of wet sand or by finer sand layers, or by both. The best-developed bubbly sand is found on the upper parts of fine-sand beaches with high tidal amplitudes. Well-sorted sand (sand with a small range of individual grain sizes) with little shell material and no mud also favors entrapment of air bubbles.

Often there is no surface expression for bubbly sand. However, one only has to walk on such a beach to know one has found *soft sand* because every step will take effort: First the foot sinks as surface pressure causes the holes and pores to compress and collapse, and then the foot must be pulled out of the compressed sand.

Most bubbly sand extends only through the surface layers, a few inches into the beach, but we have observed bubbly sand to extend as much as 1.5 feet (more than 45 cm) below the surface on the beach at Sapelo Island, in the U.S. state of Georgia, a depth that will conquer most four-wheel-drive vehicles. Driving a vehicle onto such a surface will result in getting hung up to the axles, as we have learned.

The spherical to flattened-oval holes that make up bubbly sand can be seen in a wedge of sand just extracted from the beach, or on the wall of a carefully excavated trench. The cavities range in size from 0.1 to 0.98 in (3 to 25 mm), giving the beach a Swiss cheese appearance. Near the beach surface the holes are spherical, but they become flattened somewhat due to the weight of the sand, which increases with depth from the surface. When pressure is applied, the cavities collapse, resulting in the sand's softness.

The bubbles in soft sands probably form as the result of two different processes: A breaking wave and swash mix sediment and air, trapping the air, or, more often, the rising tide forces air up through the beach. In the latter case, the lack of permeability in the fine wet sand at the surface of the beach does not permit the air to escape uniformly, but instead forces bubbles to form. There is more to this story than meets the eye, however. Many aspects of soft sand remain unstudied, and there is no apparent reason why the bubbly sands form in some places but not in others. Adding to the mystery is the bubbly sand found along some sandy lakeshores above normal lake level. Scientists assume that these sands became bubbly during local storms.

Another puzzling feature is the presence of bubbles high on the sides of now cemented dunes in the Bahamas. These bubbles appear to be identical in shape and size to those that appear in the bubbly sand on beaches. Again a storm explanation is called upon, but no one knows how they came to be. Bubbles also are entrapped in some muddy sediments, but these form from gases generated by the decay of organic matter.

Bubbly sand has a couple of practical ramifications. One is the problem created for those who drive vehicles on beaches, but in our view such driving is something that should be discouraged on most beaches anyway. It is also considered a minor problem for tourist beaches. A much more serious problem is that spilled oil and other liquid pollutants will penetrate farther into bubbly

These deep footprints are a sure sign of soft sand!

Bubble-sand layers can be buried, as seen in this trench. Note the bubble-sand layer at the base, then another layer about midway up, and finally a poorly developed bubble-sand layer near the top. This variation reflects the different conditions under which the different sand layers form, as air is trapped in the beach in repeated cycles. The U.S. dime is shown for scale.

sand (see chapter 12) than into other sands of the same grain size. Fine-sand beaches generally allow little penetration of spilled oil, so cleanup is relatively simple and involves little loss of beach sand. Bubbly sand beaches, however, can turn a disastrous spill into a catastrophic spill as far as beach life is concerned.

AIR ESCAPE STRUCTURES

Escaping air forms a large number of features at the surface of the upper beach. The names we have given these structures are not "official" names, but they are descriptive and recognizable by just about any beach stroller. One of the most common structures found on all beaches are *nail holes* (sand holes), so called because of their resemblance to a downward-tapering hole produced by a nail; even their variation in diameter and length is reminiscent of nail holes. Their abundance varies widely. By actual count on the upper portions of some southeast U.S. beaches, the density of hole spacing ranges from less than 1 per square foot to 200 per square foot (about 10 per square meter to about 2,200 per square meter), almost always on the upper beach near the highest reach of the swash at high tide.

The escaping air associated with these holes can be seen on a rising tide in the upper swash zone when the thin layer of swash water washes across the sand. As air escapes into the water and the bubbles are moved up and down in the upwash and backwash of the swash, *bubble trains* are formed. Sometimes, in response to a foot stomp, closely spaced nail holes in a dry part of the beach, away from the footprint, will collapse, forming micropits, suggesting that there is some sort of connection between the holes. Most beach strollers probably assume that such holes are produced by small organisms. Occasionally one does see sand fleas popping down these holes, but we believe they are usually occupying previously formed air holes (see chapter 9). Organisms such as sand hoppers do excavate holes that look similar at the surface, but they are typically much deeper and much less abundant than the nail holes.

If the stream of ejected air is a little stronger, it may remove sand from around the lip of the hole, making a *pit*. Pits also form where blisters collapse (see later in this chapter). If sand also is carried upward through the escape hole and deposited immediately, a *sand volcano* may form.

Where air or water moving up to the surface cannot escape through an uppermost sand layer that happens to be cohesive enough to form a trap, a cavity will form. These air or water pockets will push the overlying sand layer up into a circular dome, or *blister*, at the surface, a feature much like a soufflé. The trapping upper layer is usually fine sand, but salt-cemented sand (salcrete) or thin organic films of algae or diatoms may bind the sand to form the barrier that prevents air escape as well. Blisters often occur in clusters or fields, evenly and symmetrically spaced and of nearly uniform size. Such fields may be tens of feet (several meters) in length and a few feet (a meter or two) in width, reflecting a physical pattern of air buildup within the upper beach. Individual blisters sometimes have a nail hole in them, suggesting that a point was reached at which there was enough air pressure to create an escape hole.

Blisters can burst, leaving an irregular circular patch of different-colored sand (this feature is difficult to identify if you don't actually see the blister form and pop). Alternatively, the blister's cap can collapse or, more commonly, be decapitated and eroded off by the next cycle of swash, forming a *pit*, which is a depression that varies in size depending on the size of the original blister. More often, the truncated blisters and pits are smoothed over by the swash and backwash, leaving a *ring structure*.

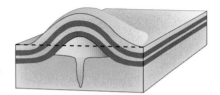

Ring structures are formed almost exclusively on beaches with a high heavy-mineral content. Alternate light- and dark-colored laminae produce the rings after the laminae are domed up by air (to create blisters) and then planed off, leaving the different-colored rings. In fact, we have never seen ring structures on carbonate beaches or beaches with very low heavy-mineral content, such as those along most of the Florida peninsula. Such rings are perhaps the most spectacular of the multitude of air escape structures on beaches.

Model of the formation of a blister and its truncation to form a ring pattern on the beach. Air and water in beach sand may be trapped by a less permeable layer of surface sand, allowing a blister cavity to form. The blister may collapse or be truncated by the next swash to form a pit. If there are laminae that are distinctly different (e.g., colored heavy-mineral layers), a ring pattern will result. Drawing by Charles Pilkey.

Clearly, an infinite variety of escape structures can be formed in a multitude of ways. We've just scratched the surface of this interesting and little-recognized phenomenon, and some of our explanations belong in the category of educated guesses. There are lots of strange things to be explained. For example, if one puts a stick or a board (perhaps obtained from the beach wrack) on the uppermost beach, the next swash that rolls over this obstruction may form a line of holes on its landward side. The edges of tire tracks and footprints likewise produce lines of nail holes and pits and may be associated with the formation of some blisters.

For the scientifically inclined beach watcher interested in the minutiae of nature, the opportunities for discovery and explanation are boundless. One might try to explain by observation how the features form, why they form in some places and not others, why they are abundant on some beaches and not on others, why they are present before or after storms, and so forth.

WATERY BEACHES AND WATER ESCAPE STRUCTURES

Water, like air, is forced in and out of the beach as the tide goes up and down. If sea level were static, the water table under the beach would be slightly higher than

Upper Nail holes (sand holes) vary in size and depth and in the concentration of their distribution. These holes are the most common feature found on beaches. The U.S. dime is shown for scale.

Center left A sand volcano produced by escaping water bringing sand to the surface. The finer, lighter-gray subsurface sand makes the tiny "volcano" stand out against the darker surface sand, with trails of the lighter sand going away like miniature lava flows. To the right of the volcano are similar trails of lighter sand from an opening that was eroded to produce a pit. The U.S. penny is shown for scale.

Lower left A blister forms as a cavity of trapped air that produces a raised surface. Note the pit to the left of the blister where a similar cavity was truncated. Nail holes are also visible. The Brazilian ten-centavo coin is shown for scale.

Center right A cross section through a blister showing the internal cavity, or soufflé structure. Note other blisters, pits, and air and water escape holes. The Brazilian ten-centavo coin is shown for scale.

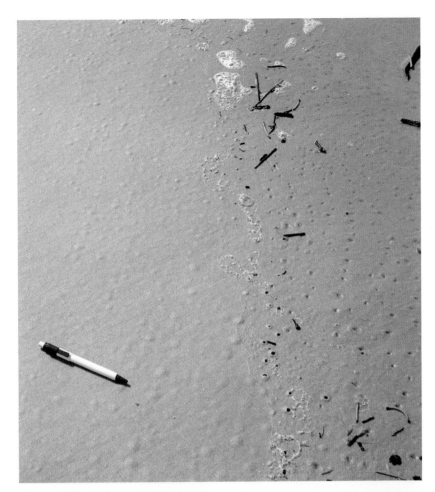

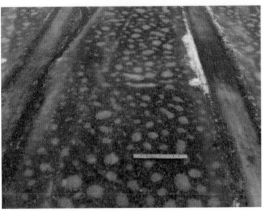

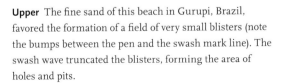

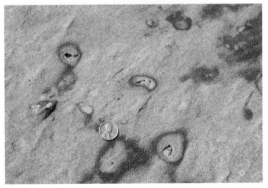

Upper The fine sand of this beach in Gurupi, Brazil, favored the formation of a field of very small blisters (note the bumps between the pen and the swash mark line). The swash wave truncated the blisters, forming the area of holes and pits.

Lower left An analogous setting for understanding how blisters form and occur in semiregular distribution patterns is seen on this Michigan parking lot in winter. Trapped air and water form and fill cavities (light areas)

between the impermeable asphalt and overlying ice layer, causing the equivalent of blisters. As on beaches, the blisters have a somewhat equant distribution. The 1 ft (30 cm) ruler and tire tracks are shown for scale.

Lower right Ring structures are produced when blister doming occurs and then is truncated by swash waves. Here the dark heavy-mineral laminae make the ring patterns more obvious. Note that the cavities left by the blisters can be seen in several of the ring structures. The U.S. penny is shown for scale.

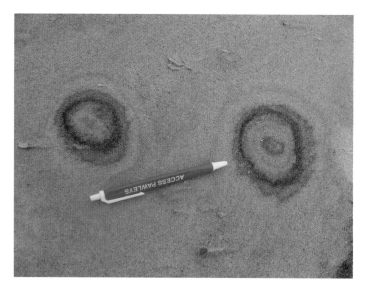

Close-up of rings.

This irregular beach surface is the result of footprints that were washed over by wave swash. Note the nail holes, larger air escape holes, and probable blisters. Sand surfaces disturbed by footprints and tire tracks seem to favor air and water entrapment, resulting in more holes. The U.S. penny is shown for scale.

sea level but would slope toward the sea to reach the same level at the water's edge. Any rise in sea level (or lake level) will flood the beach (e.g., high tide, storms, wave run-up), forcing water into the beach and raising the water table, as well as saturating or partially saturating the sand above the water table. When the tide falls or the storm subsides and the water level goes back down, the direction of flow will reverse and water will escape from the beach. This flow may take the form of springs, seeps, or even water emerging from surface holes in the beach (e.g., nail holes). On the rising tide, the bellows effect even may be strong enough to form small, temporary fountains of water a few inches high. Similarly, rainfall and surface runoff on the land may soak into sands contiguous with the beach (e.g., adjacent dunes) and then emerge on the beach as springs or seeps.

Water seepage from the beach face is common at low tide, and a seepage face develops where the water table and the beach surface contact each other. The associated features that form along this line are distinct: *Rill marks* are miniature seepage channels that commonly show *dendritic* (treelike) patterns at their upper end, braided patterns in the downflow direction, and associated depositional tongues, or microdeltas, at the terminus of sand transport. Looking down on rill marks is like looking out of an airplane flying over a river system with tributaries flowing from canyons and joining to form a trunk stream, which then turns into a braided stream. The variety of forms of rill marks is reflected by the many descriptive names given to these features: branching, fringy, meandering, conical, comb-shaped, comma-shaped, and tooth-shaped. Rill marks are found more often where there is some slope to the beach at low tide. These structures rarely form on the upper intertidal area. In contrast, the surface of flat beaches may be at the same elevation as the top of the water table at low tide, producing a *water-saturated beach* surface, which gives a mirrorlike effect. In essence, the entire beach surface is a wide seep.

Rill marks often appear at the same elevation along a beach for a long distance. This line of marks can be an expression of the top of the water table or may be the result of an invisible, impermeable sediment layer that forces the water in the beach to move out onto the beach. Peat layers often are the cause. As noted in part I, barrier beaches move landward as a result of rising sea level, often moving over and burying salt-marsh peat that existed at the back of the beach or barrier island. Such buried salt-marsh peat layers, which sometimes appear in the intertidal zone, were covered and compacted by beach

Upper A spring from which enough water is flowing to produce a small channel across the beach. Transported sand is deposited in a deltalike splay deposit in the foreground.

Lower A close-up of the spring, where water is seeping from the beach around the depression (small rills, each with a miniature depression, can be seen around the rim of the larger spring) and flowing from the bottom of the main pit. Note the three spots where the sand is being roiled up by the water outflow.

sand for as long as several thousand years before reemerging. The peat layers are impermeable, so groundwater piles up on the peat's surface. As a result, rill marks often emerge in a line on the beach, just above the peat layer. Sometimes the peat layer is visible, but if it is not, it can often be observed by digging down below the surface a little bit.

Five fountains where water and air are escaping from the beach during a rising tide. Fountains were observed forming on the berm adjacent to the sea's edge as well as in the first trough (runnel); the five shown here were under the water that had flooded the trough. Most of the fountains were short-lived, bubbling only for a few seconds, but some were intermittent, bubbling, then stopping, then bubbling again.

As noted previously, smaller water escape structures include nail holes as well as sand volcanoes, which form when escaping water from a sand hole carries sand to the surface, where it accumulates around the hole, like volcanic ash around a crater. Again, these are miniature structures, about 1 in (about 2.5 cm) in diameter, although larger spring mounds with a central pit or craterlike structure are reported in some beach sediments.

There are a large variety of other organic structures that can be mistaken for physical structures. For example, some sand volcanoes are probably organic in origin. From their burrows, ghost shrimp (*Callianassa* sp.) extrude mixtures of sand, water, and fecal material, which accumulate in the same conical fashion as the physically formed structure does (see chapter 9). Certain worms that live in the lower beach also extrude conical piles of muddy sand. As a general rule, the surface structures on the uppermost beach are entirely physical water and air escape features, but on the lower beach and tidal flats, features formed by living organisms may be present.

This chapter has brushed by a whole army of features that can be seen on the surface of just about any beach in the world. These features will be different on every

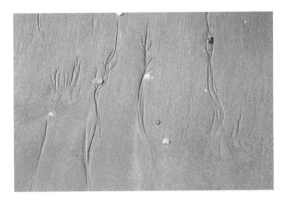

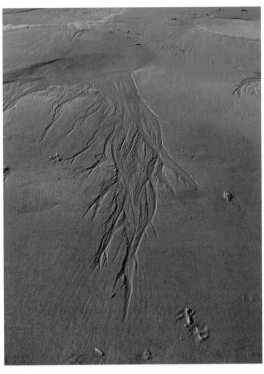

Left Rill marks form where groundwater seeps out of the beach, producing unique and varied patterns. The upper ends, or heads, of the rills are tiny dendritic, or branching, channels formed by erosion from the seeps. These miniature channels converge into main channels that terminate in deltalike splays of sand. Note the footprint to the left. The seeps have produced microdeltas that are covering the toe of the footprint, and a similar deposit can be seen at the end of the rill on the extreme right. A backwash lineation pattern is present between the rills. The U.S. penny is shown for scale.

Right This large water-seepage feature produced by converging rills at low tide on a beach in Northern Ireland is eroding headward (toward the bottom of the photo), and transporting sand seaward, filling in the trough (at the top of the photo).

When the lower beach is saturated with water, it often has a mirrorlike, or reflective, surface, as seen in this photo of the famous Atlantic City beach in the U.S. state of New Jersey. The condition is common at low tide between the edge of the swash and the upper beach.

Sand volcanoes may form as a result of either the physical escape of water and sand or the expulsion of water and sand by burrowing organisms. The origin of the prominent sand volcano in the center of the photo is questionable; however, the smaller, similar feature to the right (above the U.S. penny, which is shown for scale) has *Callianassa* fecal pellets around the opening, indicating that it was formed by the burrower.

beach to varying degrees because of multiple parameters, such as beach orientation to wind patterns, beach grain size and sorting, sand supply, heavy-mineral content, shell content, tidal amplitude, waves, and storm frequency. Much remains to be learned about these features, and they remain a fascinating puzzle, an extra and unexpected intellectual adventure for those who are willing to look at the beach as more than a pile of sand.

8

WHICHEVER WAY THE WIND BLOWS: REWORKING THE BEACH SURFACE

Beaches are windy places, both because they face vast stretches of open water over which wind blows from distant storms, and because of the differential heating between land and water. A summer morning's visit to the beach might be met by a seaward-blowing breeze, but as the day wears on, the land surface heats up at a faster rate than the sea surface and the wind direction reverses. This effect is because the ocean absorbs heat energy to a greater depth than the beach and dune sand does, so the ocean warms more slowly than the land. The warm air over the land rises as it is heated, and cooler air rushes in off the ocean to replace it, forming the familiar sea breeze.

WIND ON WET SAND AND MUD

The effect of the wind is subtle on wet sands such as in the intertidal zone, where sand does not completely dry out, or on wet beaches after rains. To begin with, wet grains are not easily picked up by the wind because the water on their surfaces "sticks" to other grains and water. Here individual grains become free to move only upon drying, but they stick to adjacent moist surfaces as the wind begins to move part of the sand. The result is an *adhesion surface* or *adhesion ripples*, a pustular- to warty-looking surface where grains have moved but eventually adhered to the moist beach surface. These surfaces tend to remain damp because water in the pores between the newly adhered sand grains also tends to adhere to the grains due to surface tension. This surface tension produces a capillary action, that is, the pulling of the water column up against the

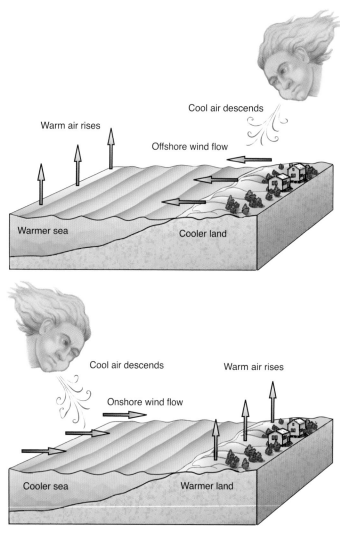

force of gravity, as can be seen when a drinking straw is placed in a glass of water. The smaller the pores, the greater this effect, so adhesion surfaces are more common in damp, fine-grained sands. As water continues to be drawn to the surface, the moist-trapping effect is maintained to accumulate more sand as it blows by. These surface features are termed *adhesion ripples* because the tiny lumps do have some asymmetry and can be used to determine wind direction, although the pattern is distinctly unlike that of the other ripple marks described in chapter 6, or of those formed by the wind in dry sand.

Such damp sand surfaces also favor the formation of tiny sand shadows, or windrows of sand accumulating in the lee of shell fragments or other tiny obstacles. These miniature *harrow structures* give a strong sense of wind direction.

Although mud is not commonly associated with sandy beach environments, here and there small patches of mud may be found. For example, mud surfaces occur on low-energy, protected

Model of the origin of sea breezes. In the morning, the air rises off of the warmer sea, causing surface breezes to blow seaward, but as the land heats up and the air begins to rise, the onshore sea breeze develops. Drawing by Charles Pilkey.

tidal flats, especially on the landward side of barrier beaches, and more rarely in beach troughs or on the uppermost beach on the back berm surface. If such a surface is covered with a thin film of water, and the wind blows over the surface, *wrinkle marks* may form in the mud. These occur as small patches of tiny ripplelike structures, only small fractions of an inch in relief. On drying, mud layers will shrink, forming *mud cracks*, which sometimes curl along the edges, causing the layer of mud to separate into thin mud chips. Such damp, muddy surfaces favor the growth of algae, and their thin mats, which give the surface a black color, help hold these mud chips together.

WET-TO-DRY TRANSITION STRUCTURES

One of the effects of beaches transitioning from a wet to a dry state is the precipitation of tiny salt crystals between the grains of porous sand laminae as the salt water evaporates from the beach. These weakly cemented sands, or *salcretes*, are more cohesive, or sticky, than dry sand and often form crusts, resistant layers that will stand in relief when the adjacent dry sand blows away. Salcretes are most common on the upper beach or in overwash passes cut through the dune line, and they sometimes contribute to a beach surface eroding differentially so that a *microtopography* develops. These irregular surfaces cut through the parallel laminae underlying the beach. If the laminae are of different compositions (e.g., alternate layers of light and heavy minerals) or different average grain size, the result is a miniature contour pattern, curious swirls of light and dark, similar to what one might see in an aerial view of a mountainous region underlain by horizontal sedimentary rock layers.

Less common structures associated with the differential drying

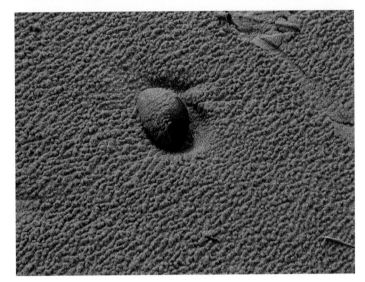

Upper Adhesion surface on a carbonate beach in the Maldive Islands. The irregular, pustular surface results from dry, windblown sand grains adhering to the damp beach surface.

Lower Adhesion surface and ripples on the beach at Portrush, Northern Ireland, including a light coating of sand on the pebble.

out of a beach are *beach biscuits*, which are found on the upper beach. The upper beach and the throats of overwash passes are flooded during storms, saturating the sand. As the sand dries out differentially, subspherical, biscuit-shaped blobs of sand remain moist and cohesive while the rest of the sand dries. The wind then removes the dry sand, leaving the biscuits standing on little mushroomlike *pedestals*. The delicate

Upper Wind-generated miniature harrow marks, or windrows, on a beach surface at Ostend, Belgium. The tiny pedestals are held up by coarser grains or shell fragments that are too coarse for the wind to move, and the fine sand streams out behind the obstacle in the direction the wind is blowing, in this case from the upper left to the lower right. The euro coin is shown for scale.

Lower Where small puddles or persistent damp areas occur at the back of the beach, thin layers of mud or algal mats will accumulate. These layers will shrink as they dry out and form classic mud cracks, or the algal mats will split and curl around the edges, forming roll-up structures as seen here on Oak Island, North Carolina. Photo courtesy of Tracy Monegan Rice.

pedestal structures tend to form after storm waves have saturated the back beach and strong winds differentially dry and sculpt the sand. The features generally do not persist, as the wind that forms them also eventually topples them.

WIND ON DRY SEDIMENT

Beaches and dunes are natural laboratories in which to see the interaction of wind and dry sand. Given its greater density, water can pick up and carry much larger material than can wind at any given velocity. Turbulence in wind, however, allows grains as coarse as sand to be picked up, and wind velocity determines how far a grain will be carried before settling to the ground. Wind is nature's best sorting agent.

Where the wind erodes and transports sands, gravel-size materials and shell hash (gravel made up entirely of shells and shell fragments) are left behind as *lag deposits*, forming a coarse, pavementlike surface that tends to retard further erosion. Fine materials such as silt- and clay-size particles may be completely removed from the beach-dune system and may be a contributing factor to the haze (which is mostly salt spray) sometimes seen down the length of the beach (and to the fine grit you can't seem to get out of your mouth, hair, or eyes at the end of the day). Sand-size particles are further sorted because the wind transports coarse and fine sand differently, *rolling* or *sliding* coarser sand particles, and moving the finer sand grains by *saltation*, a sort of bouncing or hop, skip, and jump action. The result is a concentration of well-sorted sand.

Any obstacle to the wind will increase turbulence (a chaotic air motion that allows the grains to be picked up) around the obstacle but reduce velocity on the downwind side. The result is scour around the obstacle and sand deposition in the lee of the obstacle, sometimes giving birth to an *embryonic dune*. These *wind shadow* deposits are common along wrack lines, behind clumps of plants, and downdrift of large pieces of flotsam; they even form small windrows of sand where shells interrupt wind flow. The scour around the obstacle is analogous to the scour created by wave swash around a pebble on the beach and the formation of crescent marks by backwash. Clumps of pioneer plants and dune grass have the same effect, trapping sand and initiating the formation of embryonic dunes.

Some of the first bedforms

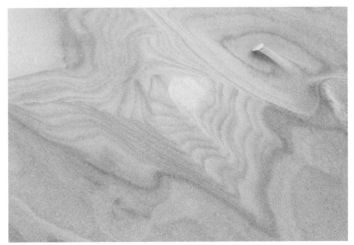

Upper Salcrete results when sand is weakly cemented by salt crystals precipitated between the grains, which happens when seawater evaporates. This layer of heavy-mineral-rich salcrete forms a crust over the quartz-rich lighter-colored sand. The brownish grains are the mineral garnet. The U.S. penny is shown for scale.

Lower Where the wind sculpts the beach by irregular down-cutting erosion through layers of different-colored sands, patterns of contour lines emerge that mimic the lines of a topographic map. Note that the shell presents a wind barrier; its presence results in a streamlined pattern. The wind appears to be blowing from the upper left.

produced by the wind are the familiar ripple marks. *Wind ripples* are parallel crested, are often split along their length, and occur in fields (i.e., extensive patches of ripples), but they differ from water-wave ripples in that the ripple crests are not as sharp and the amplitude of the ripples is less. Another noticeable difference is that the sand along the crest of the wind ripple is

Beach biscuits is the informal name used for these flat, disc-shaped clots of sand that can be seen forming from a surface layer of darker, cohesive sand (upper left) as it is eroded by the wind. The semiresistant layer may be either salcrete or damp sand, in contrast to the lighter-color underlying sand, which is noncohesive and easily blown away by the wind. The biscuits then form resistant caps that contribute to pedestal formation.

coarser than that in the trough, the opposite of the distribution in a water ripple. This difference is because the wind cannot move these coarser sand particles by saltation, so they concentrate on the crest. On a water-laid ripple, the coarser grains are carried over the crest or avalanche into the trough, where they are partially protected from the current and are concentrated. In fact, studies have shown that where the wind is energetic enough to move all sand sizes, the average grain size of the sand making up a patch of wind ripples in their entirety is coarser than the sand underlying the rippled layer. Similarly, where the wind is strong enough only to move very fine sand, a coarser lag will be left behind and the finer sand will accumulate as very well-sorted deposits in wind shadows or when the wind velocity dies away. Similarly, the wind may concentrate sand grains of different densities (placers of different mineral compositions), giving rise to color stripes between ripples. Even when ripple marks do not form, thin placer concentrations of heavy minerals give rise to black sand surfaces.

Fields of wind ripples are common on the upper dry beach, at the toe of the sand dunes, and on the bare surfaces of the dunes themselves. Often the crests are perpendicular to the shoreline, indicating formation by winds blowing parallel to the shore. The crest orientation may change in a matter of an hour or less if the wind changes direction. Few beach visitors venture out in strong winds, and those who do experience sand transport in the little stings and pinprick sensations made by windblown sand grains when they strike the skin. However, someone who can tolerate even a moderate wind can put his or her face down close to the sand surface and observe the way in which sand moves (rolling and sliding vs. saltation), and can easily understand ripple migration.

DUNES AND DUNE STRUCTURES

The recipe for sand dunes requires only the addition of a sand-trapping mechanism at the back of the beach to the existing wind and sand mix. Healthy sandy beaches are a reservoir of sand for the formation of sand dunes, and beaches and dunes are commonly associated in the beach equilibrium model (see chapter 4). Vegetation becomes important as both a sand trapper and a sand binder in the formation of dunes, although significant coastal dune fields form along shores (with large sand supplies) lacking vegetation. In the latter areas, any obstacle (e.g., large wrack objects, rock outcrops) may be the reason dune formation begins, and even where such objects are absent, the wind cannot carry sand inland indefinitely. Simple friction will cause the wind velocity to decrease and the wind to drop its sediment load.

On the dry upper beach, the small embryonic dunes that form behind large pieces of flotsam or around clumps of vegetation will grow and coalesce into a dune (*shadow dune*), and such dunes in turn will join to form sand ridges, or *foredunes* (*transverse dunes*), at the back of the beach.

The pioneer plants that sprout on the upper beach are salt tolerant and can survive the daily salt spray as well as occasional tidal flooding by seawater. Many also fix nitrogen, or remove that gas from the air to produce the soil nutrient nitrate. Initial dune formation is enhanced by dune-grass colonization: Stands of this tall grass trap sand and anchor the deposit by putting down deep

Pedestals form when a cohesive sand layer such as salcrete, a biscuit, or a shell resists sand removal while the noncohesive underlying sand is eroded away. The resulting pedestals may resemble mushrooms, little hats, or sand columns capped by a shell, pebble, or small piece of drift material. These pedestals were found on the eastern shore of Lake Michigan. Photo courtesy of Larry Fegel.

Coarse lag deposits are left behind as the wind selectively removes sand and finer sediment. Here on Mexico's Baja peninsula, the lag consists of gravel and shells, and the adjacent dunes are well-sorted sand. The surfaces of the dunes are covered with wind ripple marks.

As the wind moves sand to the back of the beach, sea oats colonies take hold and trap sand into small embryonic dunes. The increased elevation of the dune allows a creeping vine to grow and trap more sand, which will contribute to increasing the size of the dune. Note the beautifully wind-rippled surface.

Wind ripple marks tend to be lower in amplitude than water-laid ripples, and the crests are not as sharp. These wind ripples on the back of a Bogue Banks, North Carolina, beach developed in a heavy-mineral sand, and the grains are sorted by differences in their weight or specific gravity, with the red wine–colored garnet on the steep faces, the black magnetite on the lee faces, and the light mineral quartz in the troughs. The 35 mm film canister is shown for scale.

Three merging shadow dunes are in the foreground and a developing foredune is in the background. The importance of plants in trapping and stabilizing the dunes is apparent.

roots to the water table, as well as horizontal roots (rhizomes), which further anchor the sand. The laterally spreading rhizomes allow the plants to sprout shoots and spread rapidly over wide areas.

Dune grasses are most important in the early stages of foredune formation and in part determine the character of the dune. From a global view, three dune grasses are common; two species of marram grass (*mare* is Latin for "sea," so marram grasses are seaside grasses)—European beach grass (*Ammophila arenaria*) and American beach grass (*Ammophila breviligulata*)—and sea oats (*Uniola paniculata*). European beach grass is native to Europe and northwest Africa, but early in the history of British colonization this grass was spread throughout the world as a planting to stabilize sand dunes and encourage dune growth. Similarly, American beach grass is native to northeastern North America's Atlantic Coast and the Great Lakes, but the grass has been introduced to America's Pacific Coast and other parts of the world. These artificial introductions of grasses have altered the character of dunes in these locales far beyond the natural range of the grasses, to the point that some communities are now trying to eradicate such plants and return shore habitats to the original native species. Sea oats is a grass native to the southeastern United States, Mexico, and the Caribbean. Artificial plantings of this grass also are used to stabilize dunes.

While these three grasses are dominant dune formers, there are numerous other plants important to coastal dune formation and evolution, too extensive to review here. However, in nearly every beach and dune system and associated barrens, as on barrier islands, alien species are not uncommon. Another example is the Asiatic sand sedge (*Carex kobomugi*), which is native to the Pacific Coast of Japan, Korea, China, and Russia, where it contributes to sand trapping and dune formation. The plant was used as packing material in earlier times and found its way to the New Jersey coast of North America by 1929. It was found to be a dune stabilizer, and artificial plantings of this species in the 1960s and 1970s spread to become a threat to native plants along the eastern seaboard of North America,

SOUNDS OF THE SAND

Some beach and dune sands are sonorous; that is, they generate a unique sound when walked on, or, in the case of sand dunes, when winds blow over the dunes or the dune's steep-side slip face avalanches and the sand flows down. High-pitched sounds generated by beach sands are described as singing, barking, squeaking, chirping, whistling, yelping, and even burping or the croaking of a frog. The low-frequency sounds generated by avalanching dune sands are termed booming, musical, roaring, or shouting, but the sound of approaching low-flying propeller-driven airplanes comes closest to describing the roar.

An extensive scientific literature has reported on the frequencies of the sounds, the locations, and the characteristics of the squeaking sands. Differences of opinion remain as to the exact causes of the sound, although it is generally attributed to the shear that occurs between sand grains when they rotate past each other on compression, or when one layer of sand grains slides over the underlying layer. Walking or shuffling on the dry beach will cause such shear, and on the Lake Superior beach at Bete Gris, Michigan, the sound can be produced by rotating the palm of one's hand on the beach. Most singing sands will not sing once removed from the beach, yet others can be collected and placed in a beaker and will squeak when a pestle is pushed up and down in the sample (here again the sound is attributed to the effect of shearing between quartz grains).

Generally, squeaky sands are siliceous, dominated by the mineral quartz, have been subjected to wetting and then drying, and are at a low humidity. These sands are free of impurities and finer-grained material; adding dust or powder-size material to the sand will eliminate the sound, and grains with an organic coating lose their voice as well. The loss of sound production in some formerly squeaky beaches in Japan and Brazil is attributed to pollution. In other words, only "squeaky-clean" sands squeak. When naturally squeaky beaches are artificially nourished, they lose their sound characteristic.

Not too many years ago, the only way to experience such sand sounds was to visit a beach with the right characteristics and then to recognize what was happening when hearing such sounds. Today, one can find video recordings of the sounds posted on the Internet. Just Google "singing or squeaking sand" or go to YouTube and search for "musical sand dunes" and follow the various links. Squeaky Beach, in Victoria, Australia, seems to be the most popular posting, but similar sounds may be heard on Prince Edward Island, Canada, a North Sea beach in Holland, a beach in Phuket, Thailand, and South Island, New Zealand. Other famous singing beaches include Singing Beach at Manchester-by-the-Sea,

Massachusetts, beach sands on the Isle of Eigg, Scotland, and beaches on the Freycinet Peninsula, Tasmania. Virtually every continent has singing beaches.

Of particular interest are the sand sounds described as "barking," which can be heard on Barking Sands Beach, Kauai, in the Hawaiian Islands, and "whistling," which can be heard on the sands on Porthoer Beach, Llyn Peninsula, North Wales. At least thirty-three beaches in Japan have been identified as having singing sands, and resorts with the name "Singing Sands" in places such as Placencia, Belize, and Eleuthera Island, Bahamas, imply that the beaches sing there too.

The low-pitched booming or roaring sounds associated with large dunes are found both in large coastal dunes and in large inland dune fields in arid deserts. The back-beach dunes near Mana, Kauai, Hawaii, are a good example, but other large coastal dunes likely to produce such sounds are found on the Chilean Pacific Coast and in West Africa and Australia. Marco Polo reported hearing booming dune sands in the deserts of China.

particularly in New Jersey's coastal dunes. Other species of this same genus, *Carex*, are found in coastal dunes of other parts of the world (e.g., *Carex maritima* in Great Britain; *C. houghtoniana* in Lake Ontario dunes; *C. macrocephala* along North Pacific shores).

The spreading or invasion of European beach grass to the Oregon coast of the United States has created big changes in the appearance of coastal dunes and in the ecology of the beachfront zone. The native grass (*Elymus mollis*) produced low-lying dynamic dunes with overwash gaps. European beach grass created stable high dunes without gaps. The resulting cutoff of overwash sediment to areas behind the dunes created a new ecological system.

Because the marram grasses (European and American beach grasses) grow in dense stands and spread laterally in a more or less continuous pattern, the foredunes that result tend to be continuous ridges without natural breaks in the dune line. In contrast, sea oats tend to be patchy in their distribution, so where they control dune formation, the dune line is more hummocky, broken by topographic lows or swales that are natural pathways for the formation of overwash passes in storms. The same is true for arid environments, where the patchy cover of vegetation allows blowouts and breaches to form, particularly during storms. A beach with a continuous dune line at the back (typical of dunes formed by marram grasses) responds quite differently to storms than a hummocky, broken dune line (typical of dunes formed by sea oats).

The breaks funnel wind currents and may account for the natural formation of *blowouts*, bowl-shaped depressions eroded by the wind. Blowouts are the largest wind-formed erosional features in dunes. Those formed in temperate climates bottom out at the top of the groundwater table, forming a flat floor. Many of the blowouts in coastal dunes form as the result of dune destabilization by human activities such as paths from foot traffic to the beach, or vegetation kills from campfires. Countries such as the Netherlands and Denmark, which understand the importance of sand dunes as reservoirs for maintaining beaches, are strict about keeping the public off the dunes.

This foredune on Culatra Island, Portugal, in a semiarid environment, has only a sparse cover of dune grass and is subject to the formation of blowouts and breaches during storms, allowing washover to carry sand to the back side of the island. In arid regions or where sand supply is very high, the coastal dunes may be barren of vegetation and subject to rapid migration.

The eroded sides of blowouts often expose the natural internal structure of a sand dune, revealing the common *dune cross bedding*. As a sand dune grows, the sand grains are carried up the windward (back) side of the dune to the crest and then blow off into the leeward wind shadow and settle out through the quiet air; alternatively, they roll or slide down the steep leeward face of the dune. The result is that the dune face is burying sand layers, or laminae, that are inclined at angles as steep as the angle of repose of dry sand (about 30 degrees). This mode of sand movement is also how the dune migrates as an asymmetric landform. Random sections cut through the dunes will show the buried layers as cross beds at various degrees of inclination, up to 30 degrees. On the upper parts of sand dunes, where the wind has beveled off the dune and produced somewhat flattened surfaces, the exposed edges of these inclined layers also create curious patterns.

By their very nature, sand dunes tend to be unstable, cycling through deposition, erosion, and redeposition, and migration, stabilization, and destabilization. Most areas of coastal dunes have good examples of dunes migrating into or over the static structures of humans. Most coastal dunes in temperate climates are vegetated dunes and are irregular in shape. Where blowouts are common in foredunes, the resulting dune type is the *parabolic dune*, a scoop-shaped dune that is convex in the downwind direction. Such dunes form when sand supply is high or when vegetation is ineffective in trapping sand. A parabolic dune is a good place to look for dune migration features. Parabolic dune faces may show sand slump, slide, and flow features, and the landward side

Erosional bluff faces cut in dunes at the back of the beach often expose cross beds. Layers in this exposure have cross bed sets that are inclined in different directions, reflecting variable wind directions. Such exposures usually do not persist, because of slumping and slope collapse.

The slip face of a landward-migrating dune in a coastal dune field in the U.S. state of Georgia. The dune migrates as sand avalanches down the face by flowing (the tongue-shaped pattern) or by slumping or sliding (the sheet pattern), or as sand blows off the crest of the dune.

of the dune shows evidence of burial of vegetation or whatever is in its migratory path.

Like beaches, dunes are strongly influenced by sand supply. In some cases, dunes are very ephemeral, reflecting a relatively short-term event that led to dune formation but then ceased. A good example is when a tidal inlet closes. The offshore ebb delta in front of the inlet is a reservoir of sand. When the inlet closes, much of that sand may come ashore, feeding shoreline accretion and dune formation. Once the reservoir is exhausted and the offshore body of sand is gone, the pattern changes to shoreline retreat, and dune growth either ceases or is greatly reduced. However, where there is a healthy and continuous sand supply, great dune fields form. Some noteworthy examples include Coos Bay, Oregon, in the United States; the Brazilian Gurupi Coast, south of the Amazon; Britain's Sefton Dunes; France's Bay of Biscay coast and the Great Dune of Pilat, Europe's largest sand dune; the dunes of the Curonian Spit in Lithuania; the Doñana National Park on the Gulf of Cadiz, Spain; in Africa, the Morocco Atlantic Coast, the Alexandria Coastal Dunefield, and the Namib Sand Sea of Namibia; in Asia, the Tottori dunes of Japan and Mui Ne in Vietnam; and in Australia, the Cronulla sand dunes near Sydney, Mount Tempest near Brisbane (Australia's highest), and Fraser Island, Queensland.

The combination of high winds and blowing sand produces another interesting artifact of nature found both in deserts and coastal dune fields. Pebbles and cobbles exposed to the wind are sculpted by sandblasting to form smooth faces that intersect each other with sharp edges. If the pebble is partly buried, only the exposed portion will have the flat faces, while the buried portion remains rounded. Appropriately enough, these rocks are called *ventifacts*.

The next chapter (chapter 9) reviews some of the array of sedimentary structures that form from biological processes in dunes and beaches. In addition to wind ripples, the transition zone from the back of the dry beach into the toe of the dunes is characterized by animal traces, typically crab tracks and their burrow holes, as well as a variety of insect trails and burrow traces (e.g., from burrowing beetles, bees, and fly larvae). At

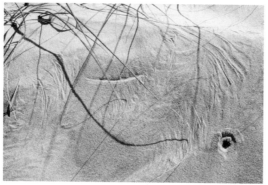

Upper left Praia do Guincho, Portugal, is famous for its prevalent onshore winds and sand dunes that migrate inland from the beach. Limestone and gravel derived from local outcrops are sandblasted by these winds to form ventifacts and outcrops with sculpted faces. The camera lens cap is shown for scale.

Lower left Scribe marks produced by the ends of dune grass fronds moved back and forth by the wind. Note also the opening of a hole produced by an unknown dune burrower.

Upper right Extensive burrow traces reflecting activity just below the surface of a dune. Different-size burrows indicate different organisms, but beetles and bees are suspects here. The U.S. nickel is shown for scale.

Lower right A trail of reptilian footprints across a wind-rippled dune surface near Colan, Peru.

the vegetation line, circular to semicircular *scribe marks* etch the sand surface where the tips of grass fronds, stalks, or fine rootlets have been dragged around by the wind. Patches of dune grass are favorite sites for ant lions to build their conical pits to trap other insects. Throughout the dune system, animal tracks, trails, and droppings can be identified. Bird footprints are the most common, but tracks of reptiles, rodents, and mammals are easily spotted on the bare sand surfaces.

DUNE PLANTS: SURVIVING IN A DESERT

The sand dunes behind a beach are a relatively arid environment even in moist climates. Pour a cup of water into dune sand; where does it go? The answer is down, all

FULGURITES

Beaches and dunes are the last places you want to be in an electrical storm. Most beach-safety lists stress water and sun hazards, but they rarely stress getting off the beach and dunes before a storm arrives. Lightning strikes on beaches are common, and fulgurites are reminders of past strikes and their intensity. Sometimes called petrified lightning, fulgurites are tubular structures of natural glass that form when silica-rich sand (e.g., quartz, feldspar) is melted by the intense heat (3,270°F [1,800°C] or greater) generated by a lightning strike. The sand melts and immediately chills to glass, and sand grains fuse to the surface of the melted sand. Although fulgurites up to 49 ft (15 m) long and several centimeters in diameter have been reported, the resulting tubular structure is usually a fairly small, glassy tube a few inches long, with a rough, irregular outer surface. The tubes are sometimes curved, rather than straight, and may branch. Similar structures can form where high-voltage electrical lines come in contact with siliceous sand, as when lines break during storms. Again, this is not a place you want to be when a fulgurite might form.

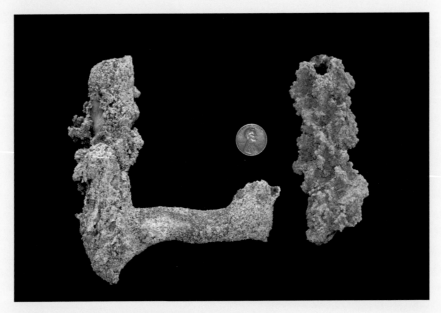

Two fragments of fulgurites showing fused sand and the characteristic hollow-tube structure. Cemented sediment fill of burrows sometimes is mistaken for fulgurites; the best way to avoid this error is to confirm that the fulgurite has glass lining the tube (at the top of the sample on the right and in the patch exposed on the upper left side of the sample on the left). The U.S. penny is shown for scale.

Collectors who seek fulgurites are most likely to find them in dunes, including those in noncoastal deserts. Cemented burrow-fill sediments are sometimes mistaken for fulgurites because of their shape, but the genuine article can be identified by the appearance of glass: smooth, translucent, and often containing small bubbles. The color ranges from dark grays to green to tan or almost colorless, depending on the initial composition of the sand.

the way until it reaches an impermeable layer such as a salt marsh peat, another impermeable deposit, or the water table. When the dunes are 100 ft (30 m) high, the plants on top of the dune crest have to go down deep to get water. Because the adaptation to produce deep roots is not common, very few plant species can survive in a sand dune field. Those that do survive have deep, water-gathering root systems.

Coastal dune plants also must tolerate salt spray and a limited supply of nutrients. Try spraying your garden plants with salt once or twice a week and not providing them with food (i.e., use dry sand instead of mulch or potting soil). Your garden would soon be barren, but dune plants are tough and have developed special adaptations to survive in the harsh coastal environment.

Everyone thinks of a cactus as a desert plant, and its defining characteristics are its needles. These are the cactus's leaves. Nature has bundled them up into sharp needles to reduce their surface area because plants give water back to the atmosphere, or transpire, at a rate proportional to the area of their leaves. There are cacti in dunes (e.g., prickly pear) for the same reason. Other dune plants simply roll up their leaves to conserve water during the hottest part of the day. In another strategy, dune plants store water to survive droughts. These succulent plants are common in deserts as well as in coastal dunes. Finally, another mechanism for dune plants to retain what little water they get is to cover their leaves and fruit with a waxy coating. The wax coating on such plants as bayberry (*Myrica* sp., used to make candles) reduces evaporation, and the ability of the roots to fix nitrogen allows the plants to live in poor soils.

A walk on a path through the dunes as you leave the beach presents a fascinating series of botanical changes that are driven by plants' differing abilities to survive in the coastal desert. On the edge of the sea, dune colonizers fix nitrogen and, along with occasional pieces of seaweed, create a rudimentary soil. These colonizers are replaced in a landward direction by dune grasses, usually a monoculture (a single plant species) that sometimes extends over a vast area. Connected by horizontal roots, or rhizomes, single living plants cover extensive areas of these dunes. This characteristic was once discovered in the U.S. state of New Jersey when an herbicide was sprayed on a small area of dunes but killed the plants beyond the spray zone. These dune plants reproduce for the most part vegetatively, that is, by having new shoots grow up from rhizomes. The advantage of this type of reproduction is that it would be a waste of energy for the frontal plant to go to flower and seed because the plants grow so densely and soil conditions are so poor that a seed would not have much of a chance for germination success. Look closely in July or August at the marram grasses and note where the plants have flowering or seed heads. They flower only at edges of the dune field, where windblown sand partly covers their stems. In fact, that is the triggering mechanism for flowering in marram grasses. When sand is free to blow, there are not many other plants around (the nearby areas are bare beach or a pathway), and it is worth the plants' effort to produce seed.

In the more landward reaches of a dune field, shrubs appear, such as seaside roses and bayberries in eastern North America. These plants need a more stable, nutrient-rich soil than the marram grasses. Marram grass is so adapted to surviving in its seaside location that it cannot outcompete even the most common weeds in a terrestrial field (any more than field weeds could ever survive in a sand dune environment). Shrubs cannot endure salt spray like marram grasses either, and so they appear only in more landward locations. Even in a more landward position, shrubs and trees are subject to the effect of the salty sea spray that is released from the bursting bubbles of breaking waves. The salt air prevents the formation of new leaves and branches on the seaward side of the vegetation zones closest to the shore, resulting in *salt pruning*, the characteristic sloping of the vegetation profile from the crown of the maritime forest down across the shrub line to the dune meadows.

In the most landward positions, the climax or ultimate sand dune plant community, the maritime forest, is found. Trees vary globally, but stunted evergreens such as pitch pines are common, as are oak, cedar, holly, and other pine, as well as a unique undergrowth community.

This change from one plant community, specially adapted to its place in the dunes, to another is called botanical succession, each group of plants succeeding the previous group. As a beach grows seaward, the pioneer plants at the back of the beach are followed by marram grasses, which are followed, or succeeded, by shrubs and then by the maritime forest. Each community prepares the way for the next, and the process leading to a maritime forest can take, literally, hundreds to thousands of years.

What does it mean, then, when we observe a maritime forest at the edge of the beach, with its trees falling into the sea? The answer is either natural or man-made change. One scenario is that the beach grew seaward during a time of slowly rising sea level and the plant-succession zones developed, leading to a maritime forest some distance in from the shore, but the rate of sea-level rise now has increased, and rapid shoreline erosion has allowed the migrating beach to erase the dunes and cut directly into the forest. Alternatively, perhaps the abundant sand supply that helped maintain a healthy dune system has been cut off by engineering structures to protect buildings, and the lack of sand supply contributed to dune loss and beach retreat into the maritime forest.

9

BEACH CREATURES: TRACKS,
TRAILS, AND TRACES

At the end of a sunny day, a popular recreational beach can resemble a plowed field, the surface churned up by thousands of human footprints. But if we saw that beach for the first time after the crowd had left, could we tell what had roughed up the surface? Up in the wrack line there might be a concentration of cigarette butts and beverage containers that would give a few clues that humans were involved, and on some wet, hard-beach stretches, footprints may be preserved.

It is not only humans that can churn up a beach surface, however. In addition to a myriad of marks and bedforms created by wind, waves, and currents, most beaches reveal that they are alive with the activity of animals that leave a variety of tracks, trails, and traces as evidence of their presence.

The traces animals and plants produce in the beach and nearby tidal flats and dunes are as fascinating and informative as the physical structures formed by waves and currents. Such tracks, trails, and burrows are seen more frequently than the animals that produced them. Perhaps we see the same birds that left their footprints, but the crab burrow holes, their pellets, and the web of trails that run from one hole to another are certainly a more common sight than the shy animals that left these traces. Lucky observers may see the morning's trail of a turtle's crawl across the beach from the night before, but fewer people have actually witnessed such a critter's beach visit. Other large creatures such as seals, walruses, and sea lions leave characteristic crawl tracks that persist until erased by a high tide or a storm. Relating these sometimes mysterious, sometimes obvious features to their seen and unseen parents is part of reading the

Crab excavation (not fecal) pellets and claw prints around a burrow opening, Cape Tribulation, Queensland, Australia. The Australian fifty-cent coin is shown for scale.

An elephant seal crawl track on an Antarctica beach with a penguin shown for scale. Photo courtesy of Norma Longo.

beach. This chapter highlights the organisms that live on and within the beach. The following chapter examines shells and skeletal material that become an important part of beach sediment.

BEACH ANIMALS FROM MICRO TO MACRO

The beach is a hostile environment in which to make a home. To live on the beach, an animal must adapt to changing wave energy, rapid changes in the shape of the beach (taking place in a matter of hours), swift and variable currents, frequent movement of the sand and gravel, alternate wetting and drying, strong winds, changing water salinity, changing temperatures, and rapid burial or exposure. Yet the beach is alive. It is not the jungle, but the place where you spread your beach towel is likely to be in the midst of millions of animals. It is true, however, that most of these animals are at or near the microscopic level, living between the sand grains of the beach, and only a few are in the size range that we are likely to see.

Biologists divide these beach animals into *microfauna*, which are of microscopic size; *meiofauna*, which are the size of small sand grains, barely visible to the naked eye; and *macrofauna*, which are big enough to be seen readily, although many of these also are very small (the size of coarse sand) and usually escape our attention. Another term, *psammon*, refers to both very small plants and animals that grow on, grow in, or move through the sand of the beach.

Widely spaced kangaroo tracks on an Australian beach.

The psammon and meiofauna generally leave no visible traces or evidence of their presence; however, the abundant presence of algae or diatoms (usually considered to be a type of plant) may give a greenish or slightly yellowish tinge to the wet sand. The meiofauna are so small that most can live between the sand grains, moving freely through the pore spaces without displacing the grains. Their abundance in the sand is often mind-boggling; words like *millions* are needed to describe population densities. Yet if you scoop up a handful of sand, you may not see a single one. Some of these organisms are phosphorescent, however, and if you stroll on the beach at night, try stomping on the surface or dragging a stick through the sand; it may stimulate the critters to glow and reveal their presence, if only briefly.

The tiny organisms of the beach are also very diverse in the numbers of different species present and the variety of roles they play in their home environment of the pores between the sand grains. Nematodes are often the most abundant, typically accounting for about 85 percent of the meiofauna. They feed on algae, bacteria, and organic detritus and serve the natural role of beach cleaning. They are, however, part of a larger food chain; the meiofauna are the food source for a range of larger animals from invertebrates to fish to birds. This dual role of cleaning the beach and being near the base of the food chain is why the abundance and diversity of the meiofauna are often used as a measure of the beach's health. Artificial beaches constructed from dredged

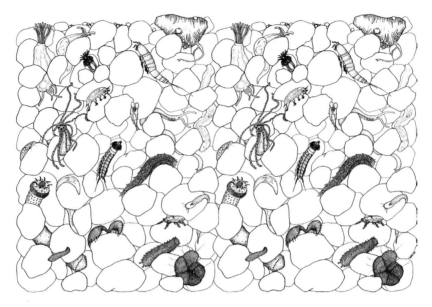

An idealized illustration of the meiofauna that live between the sand grains of a beach. Illustration courtesy of the Department of Invertebrate Zoology, National Museum of Natural History, Smithsonian Institution.

sand certainly lack such fauna in their initial construction, and beach bulldozing and daily manicuring severely impact or destroy the meiofauna, as well.

MACROFAUNAL CLUES

The beach macrofauna include the visible animals that live in the beach as well as those that visit the beach regularly, either from the land or from the sea. These creatures leave evidence of their presence and their activities in the form of footprints, tracks, trails, burrows, or fecal material.

MOLLUSKS (CLAMS AND SNAILS)

The mollusk group accounts for most of the shells and the calcareous sand on the beach (see chapter 10), but most live offshore, and the traces of clams and snails are not particularly common on the beach. Perhaps the two most globally common clams in beaches are the surf clams (*Spisula* sp.), sometimes called pipi, and the numerous species of the coquina clam (genus *Donax* sp.), also known as bean clams. These beach clams are found from temperate to tropical climatic zones. (The designation of [*Name* sp.] for an organism refers to its generic name and indicates that there is more than one species in the genus.) Commonly found in the swash zone in groups, these small, colorful clams burrow just below the surface, forming a tiny pit in the beach.

Donax clams live within the beach in a vertical orientation, obtaining food by filtering the water that the waves move back and forth. The clams protect themselves by burrowing deep within the beach when they sense that storm waves are on the way. While alive, they seem to congregate in patches, some of which may have up to ninety individuals per square foot (about 10 cm²) of beach. After their death, their shells can make up a major portion of the seashells on the beach.

The clams extend their siphons just out of the sand to form two small holes, and the holes and pits are somewhat like the nail holes produced by air escaping from the beach (see chapter 7). These micro-irregularities formed by *Donax* siphons may interfere with swash flow and generate a pattern somewhat like rhomboidal ripple marks. Breaking swash may briefly expose these clams, but they quickly burrow back into the sand. Larger clams (e.g., soft-shell clams and razor clams) may be found as burrowers in tidal flats or the beach at low tide.

Small snails sometimes leave a continuous trace on the beach surface as they move in what appears to be a random pattern. At low tide such snails may be numerous on tidal flats as they graze over the ripple marks. Some snails (e.g., the North American moon snail [*Polinices* sp.] and the lettered olive [*Olivia* sp.]) burrow just under the surface and then laterally as they feed, leaving

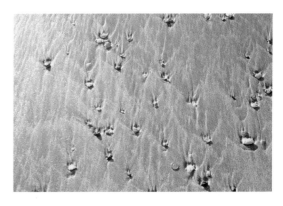

Groups of small clams living just below the beach surface become obstacles to backwash flow, creating rhomboidal and obstacle-mark patterns. A U.S. quarter is shown for scale.

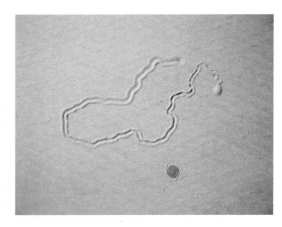

A mollusk burrow trail just below the surface of a Gold Coast, Australia, beach. The Australian twenty-cent coin is shown for scale.

a feature that looks like a sharp slash mark in the sand or a convoluted trail of disturbed sand. If you wish to find out who the trail maker is, follow the trail to its end and pluck out the shell! Just be sure to put it back in place once you've viewed it!

CRUSTACEANS

Crustacea is a subphylum of the Arthropoda, a group whose tracks are well represented on the beach. *Crabs* are the most familiar; worldwide, crabs include numerous genera that appear on beaches and their associated environments, both offshore and onshore. Their most common traces on the beach are burrow openings, footprint trails between burrow holes, feeding trails, and fecal pellets. Often these feeding

A snail's travel trail on Playa de la Griega Colunga, northern Spain, with a curious loop as if he had temporarily forgotten where he was going. The snail is at the lower right edge of the photo, near the ten-cent euro coin (shown for scale).

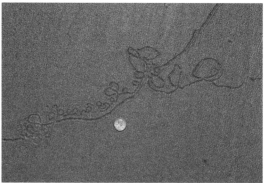

A snail trail that looks as if the creature was trying to write a message. The straight portions of the trace are travel trails, made as the snail went from one point to another, whereas the pattern of small loops is probably the animal's feeding or grazing trail. The U.S. penny is shown for scale.

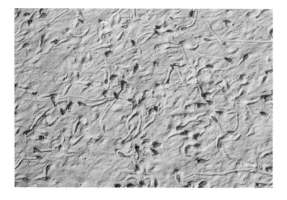

A concentration of small periwinkle snails grazing on a beach.

A moon snail beginning its burrow into the beach.

trails involve a systematic pattern of scraping food from the surface of the beach and leaving the remnants as small pellets. On the Queensland coast of Australia, beaches that are initially covered in antidunes as the tide falls become the scene of frantic feeding activity by crabs that emerge at low tide. Soon the whole beach surface can be covered in feeding traces radiating out from the central dwelling holes. Crabs vary in behavior; they may be shy creatures, such as the ghost crab (*Ocypode* sp.), or seemingly fearless crabs that gather in large groups, such as fiddler crabs (*Uca* sp.). Similarly, crabs range in color from almost colorless, to camouflage colors that blend with the sand, to vivid bluish, green, orange, and red. Crabs are common on the upper beach and in closely associated environments such as salt marshes, mangroves, and mudflats (e.g., fiddler crabs, with their one large fiddlelike claw, and hermit crabs make these their home). Their burrow structures can be complex and may extend for several feet below the surface of the beach.

Sand hoppers, which belong to the groups known as amphipods and isopods, are also common on many beaches along the wrack line or just at the high-tide line. These small nocturnal crustaceans move onto the wet beach at night to feed and by day retreat up the beach to burrow into the sand or use the shade of wrack to protect them from the heat of the sun. Large groups of hoppers may be found beneath piles of sea-

A crab feeding trail and fecal pellets pattern, Hillsborough, Australia. The Australian fifty-cent coin is shown for scale.

weed, which is one of their main food sources; they release nutrients back into the food chain or are themselves eaten by birds. Although they often dig their own burrows, sand hoppers utilize the existing air escape holes in beaches for their daytime hideouts. Where they do burrow, they may leave telltale excavation patterns on the surface of the beach. Smaller species generate burrow holes about 0.08 in (2 mm) in diameter and only 0.4 or 0.8 in (1 or 2 cm) deep, leaving the excavated sand in a tiny anthill-like mound.

Larger species, the size of jelly beans, dig their holes by powerful backward kicks that throw the sand back onto the beach, then alternately dig on each side of the hole, producing a two-rayed pattern on the beach surface. Because there are usually several of these animals living together, the abundance of holes and the characteristic ray pattern, which looks like "chicken tracks," are giveaways of their presence. These holes are elliptical in shape, on the order of about 0.5 in (5 to 12 mm) at their widest, and up to 10 in (25 cm) deep.

The greater size of these holes and the associated ray pattern make them distinguishable from holes produced by the physical escape of air or water from the beach (see chapter 7). However, the presence of the holes has been associated with the formation of sand blisters or domes. Such blisters may be initiated when the tide line rewets the burrowed beach and a wave seals the top of the burrow with wet sand. Water seeping into the beach forces air up the former burrow hole, but the wet sand at the surface of the beach effectively traps the air, which then lifts the overlying sand layer into a blister. As with other similar blisters, these can be truncated by waves to form pits and rings (see chapter 7).

Upper A mysterious pattern of mounds on a beach in Marbat, Oman.

Center A close examination reveals that the mounds are the sand excavated by a type of ghost crab from their burrows. This close-up of one of the pinnacled excavation mounds associated with a burrow opening shows the crab's fecal pellets on the front side of the mound and its claw traces from the burrow to the right side of the mound. Photos courtesy of Miles Hayes.

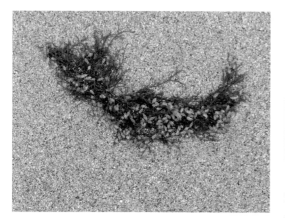

A dense cluster of sand hoppers feeding on seaweed flotsam.

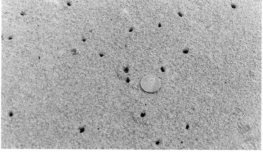

Sand-hopper holes on a Japanese beach. Note that the holes are irregular in diameter, unlike many nail holes and air holes. Some of the openings are surrounded by a rim of excavated sand. The U.S. penny is shown for scale.

Similar visitors from nearby dunes or other environments can come down onto the beach to feed. In South Africa, curious traces spotted on the upper beach turned out to be made by tiny crickets from the adjacent forest, feeding on the beach after dark.

Mole crabs (including such genera as *Emerita* sp. and *Mulina* sp.) are another common group of crustaceans in the beach; however, they are found in the edge of the breaking surf and lower swash zones. Although they do not leave common traces on the beach, you may see them when you are wading along the water's edge. These suspension feeders live in groups, burrowing just below the sand surface with their antennae exposed to capture zooplankton. When a breaking wave or backwash exposes the mole crabs, they quickly rebury themselves. They are fairly easy to capture as the swash returns to the sea.

Sand hoppers and mole crabs sometimes show up under other common names such as sand fleas; however, they are neither insects nor fleas. The term *sand flea* suggests a small, biting creature, and the term is commonly used for tiny mosquitoes and "no-see-ums." A true flea, the chigoe, is sometimes common on beaches in tropical and subtropical Africa and the Americas.

Another Crustacean that leaves traces on the intertidal beach is the genus *Callianassa*, which has numerous species. These shrimplike animals belong to a group known as the decapods and are deposit feeders that burrow deep into the sediment. Sometimes called mud shrimp or ghost shrimp (although other genera go by the latter name also), *Callianassa* species are found worldwide. Depending on the species, they range in habitat from intertidal beaches and mudflats to subtidal depths. Their burrow openings are on the order of 0.25 in (about 0.5 cm), and the burrows are intricate and extend to depths of as much as 6 ft (about 2 m). As they burrow, sand and water may be extruded from the burrow opening, forming miniature sand volcanoes and flows. They also extrude fecal pellets that surround the burrow opening, having the appearance of tiny cylindrical chocolate sprinkles. Because they are deposit feeders that strip microorganisms from the sediment as they burrow, they tend to prefer fine sand to muddy sediment, so the burrow holes are more likely to be found on fine sandy beaches and tidal flats at low tide.

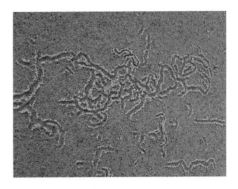

This pattern was made by sand hopper–type organisms on the beach surface, but one does not always know which specific organism is producing such a pattern. Photo courtesy of Dave Schoeman.

The forest cricket Tridactylidae comes out at night onto the adjacent beach at Cape Vidal, South Africa, leaving its traces behind. Photo courtesy of Lynette Perissinotto.

These burrows are more than just holes in the sand. The animal builds a slightly cemented wall to give strength to the burrow opening, one strong enough to resist the inflow and outflow of water during a tidal cycle. When waves or swash erode sand from around the burrow opening, the wall may be left standing as a tube or a *chimney structure*.

Callianassa major, a species common to beaches along the southeastern U.S. coast, can be found from the intertidal zone to the midtide line and no higher. This fact has been used to determine the level of the sea in fossil sands from ancient shorelines left behind by higher sea levels on coastal plains.

CHELICERATA

Horseshoe crabs are arthropods that are not crabs but are more closely related to the arachnids (e.g., scorpions and spiders) and have their own taxonomic class, the Merostomata, in the classification of the Arthropoda. A close relative of the long-extinct trilobites, which existed hundreds of millions of years ago, horseshoe crabs are said to be "living fossils," and their carcasses are commonly found washed up on beaches. During mating season they come into the shallow waters adjacent to the beach (e.g., off Delaware Bay, on the Atlantic Coast of the United States), and they are sometimes seen in association with their tracks in the sand. The group is represented by three genera: *Limulus* sp., along the North American Atlantic and Gulf of Mexico coasts; *Carcinoscorpius* sp. and *Tachypleus* sp., on the east coast of India; and species of the latter genus in some areas of the Japanese coast.

POLYCHAETES

Sharing beach space with worms does not sound too appetizing; however, the polychaetes are a class of annelid worms that includes numerous species found in nearly all intertidal zones, some leaving unique traces. Although they are too numerous to outline in detail, the following are good examples. The common lugworm (*Arenicola* sp.) lives in U-shaped burrows on tidal flats and excretes a distinctive coiled casting that has the appearance of brown paste squeezed from a tube. Numerous piles of the tubular mud mark burrow openings where the worms live in close proximity to each other. The common genus *Balanglossus* lives in similar-shaped burrows and also produces ribbons of excreted sand with the same appearance. Some of the polychaetes produce burrows that are lined with material that persists when the surrounding sand is washed away. For example, some secrete a chitinous material to paste together shell fragments and plant debris to line their tubes, which stand as little pipes when exposed by erosion. Some burrow tubes are filled with sediment and shell debris that become cemented, thus preserving the shape of the burrow, even when the burrowing organism is unknown. Sometimes tubes may remain intact when the surrounding sand is washed

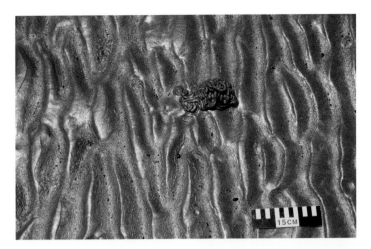

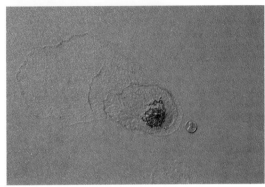

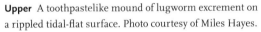

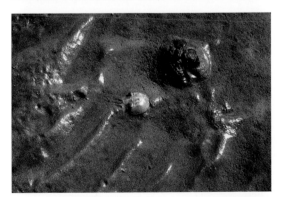

Upper A toothpastelike mound of lugworm excrement on a rippled tidal-flat surface. Photo courtesy of Miles Hayes.

Upper left A *Callianassa* sp. burrow opening from which the animal has extruded a sand-water mixture to form a sand volcano. The pattern around the little mound is where water flowed away from the opening. The U.S. nickel is shown for scale.

Upper right *Callianassa* sp. fecal pellets around a burrow opening. The pellets have a characteristic appearance of chocolate sprinkles. The U.S. penny is shown for scale.

Center left A small horseshoe crab leaving a very faint trace on a rippled tidal flat. The sun is reflecting off the ripple crests and the burrow mound to the right.

Lower Sometimes burrow tubes are filled with sediment and naturally cemented by water percolating down through the sediment preserving the shape of the burrow. This example preserves small snail shells in the fill as well. The animal that produced this burrow is unknown. Did it selectively bring the snails into its burrow, or were they simply washed into the opening at random? The U.S. quarter is shown for scale.

away, or the cemented fill is eroded and washed up on the beach. The *Onuphis* sp. lines its burrow with a parchmentlike substance that sometimes is exposed at low tide, appearing as limp tubes that feel slimy underfoot.

One way to search for these creatures that we know are there, but rarely see, is to screen the beach or tidal-flat sediment. Use a piece of window screen or a kitchen strainer and wash the beach sand or tidal-flat sediment through your sieve. You may find some very curious-looking animals.

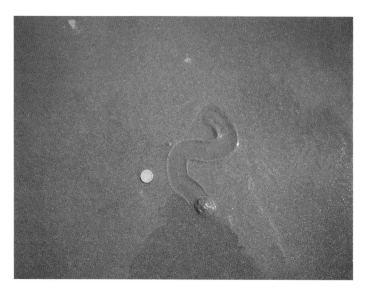

A characteristic trail of a sand dollar on South Mission Beach, Australia, at low tide. The animal has a covering of tiny spines that allow it to move—slowly—and to burrow as seen here, just beneath the sand surface. The Australian fifty-cent coin is shown for scale.

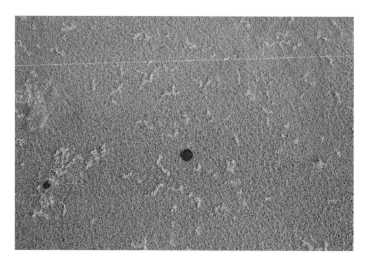

Fly larvae traces (white patches) are burrow structures just below the surface of the beach. This beach surface has a roughened appearance due to raindrop impressions. The U.S. penny is shown for scale.

OTHER INVERTEBRATE ANIMALS

Sand dollars (*Mellita* sp., *Dendraster* sp., *Echinarachnius* sp.) normally live offshore but on occasion are found at low tide, burrowing just below the beach surface and leaving a trail the width of the shell. Perhaps these are individuals that survived being washed into shallow water by storm waves.

Other burrowers include a variety of flies, beetles, and other insects or their larvae, which leave unique patterns in the sands of both beaches and associated dunes. Sometimes, when an insect burrows just below the surface, its trace may be followed by a bird that pecks at the trace in search of food, thereby enhancing the appearance of the trace. The same is true for dead creatures washed up on the beach. They can be surrounded by the trails of a variety of beach creatures that emerge to feed on them. Similar traces in the dunes also include those created by ants, beetles, and bees, as well as the characteristic semicircular to circular *scribe marks*, where the tips of grass blades or ends of plant stems etch lines in the sand as they move in the wind (see chapter 8).

VERTEBRATES

Birds are common beach visitors that come to feed on the abundant microbes and small invertebrates that inhabit the beach. The association of specific birds and their footprints is one of the more obvious examples of identifying and interpreting tracks and associating patterns with behavior (e.g., feeding or resting). Some species nest on beaches (e.g., various plovers, terns, and penguins) or in the nearby dunes, marshes, or adjacent sea cliffs. Piping plovers and least terns are listed as endangered species in the northeastern United States, and their presence has stopped beach development in Maine.

Depending on the climatic zone, several types of other large animals visit beaches and tidal flats. Sea turtles visit beaches to lay their eggs or simply to lie on the beach and bask in the sun. From the sea come seals and sea

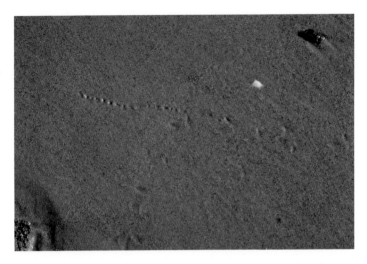

A bird's foraging trail. Indistinct bird footprints come into the photo from the right, and then the bird pecked at a food source that was just below the surface, producing a line of tiny bumps and depressions (beak marks). After lunch, the bird went on its way as the tracks show, toward the bottom of the photo.

A scribe mark etched as a perfect circle by the end of a single windblown dunegrass frond, South Stradbroke Island, Australia.

lions, which leave their trails of movement or resting troughs, while from the land come such animals as deer, horses, cattle, foxes, raccoons, rabbits, crocodiles, lizards, kangaroos, lions, leopards, camels, elephants, and smaller animals such as mice, depending on the location. Even snakes visit the beach, and especially the dunes.

A small stingray pit on a rippled tidal flat.

Sea turtles regularly come ashore on the black sands of Punalu'u Beach, Hawaii, to rest and bask on the beach. Photo courtesy of Norma Longo.

All are recognizable because of unique tracks or trails. One of the authors was surprised to see three elephants on a beach in Northern Ireland but later discovered they were from a traveling zoo and were being exercised, not unlike the Irish racehorses that are exercised on beaches. Along the shores of Maputo Bay, Mozambique, in West Africa, two of the authors saw a tree on the beach that had been knocked down by an elephant the previous day.

The trails of these and domestic animals give clues as to what other creatures use the beach. We wondered why and how cow footprints covered the beach of an isolated barrier island we visited in Turkey until we saw a herd swimming across a tidal inlet, the herder holding on to the tail of one of his swimming charges.

EVIDENCE OF THE HABITAT ROLE OF ANIMALS

Just as thousands of human footprints rework the sand surface of recreational beaches, the animals that burrow and till the sand below the surface are the natural reworkers of the beach sediment. *Bioturbation* is the term given to this natural mixing of sediment by animals. Just as the surface sedimentary structures formed by the waves and wind can be destroyed by animal traces, in the subsurface, beach bedding and laminations can be crosscut by streaks of burrow backfill or entirely eliminated as a result of the churning, reworking, and ingesting of sediment. Fecal pellets are evidence of such feeding habits altering the sediment. Other organisms, such as some snails, bore into other invertebrate shells and skeletal parts to produce sand grains. Similarly, offshore, boring animals (e.g., starfish, snails, some sponges) break down shell material into sand-size and finer particles (see chapter 10).

Upper left Bioturbation exposed in a trench at the back of a beach. Note that some horizontal laminations have been completely destroyed by lateral burrowers, while other layers show vertical crosscutting patterns of sand-filled burrows.

Upper right Marine plants anchored to cobbles and pebbles may be washed ashore during storms, rafting the attached rock to the beach. Such rafted material is a good example of the "unexpected" find in beachcombing.

Lower The bored clamshell (note numerous holes in the shell) in the center of the photo reflects another type of feeding, in which an organism attaches itself to the clam and rasps or bores a hole to get at the soft parts. Such borings help break down the strength of the shell, allowing it to be broken more easily by wave action or by turbulence during the pumping of sediment onto the beach for artificial nourishment. The U.S. penny is shown for scale.

Some beach discoveries are best left alone. Snakes frequent dunes and beaches, and this North Carolina copperhead (poisonous) has found an ideal spot to bask in the sun! Photo courtesy of Elizabeth Ray-Schroeder.

Burrowers are also natural aerators of the beach, both while burrowing and through their open burrows. Deep burrows associated with some crabs and decapods are important pathways for air and water to infiltrate well below the beach surface, providing for respiration of other organisms and creating oxidizing conditions. At the same time, such open passages can be detrimental in the case of oil spills, allowing the oil to find its way into the depths of the beach quickly.

The tracks, trails, and burrows of beaches are the undeniable evidence that a huge and highly varied ecosystem lives in this, the most dynamic environment on earth. Anyone who is curious about the plants and animals of the beach, how the beach works, and what the components are need only spend time on the beach looking at tracks and trails, digging, screening, and observing. It is a fascinating exercise for anyone at any age.

Other organisms found on the beach may not be from the beach, but rather have been washed ashore, particularly during storms. Some of these are unusual occurrences—and some are organisms that should be approached with caution and not handled! Chapter 10 includes a more extensive survey of beach organisms, and particularly those organisms whose remains may wash ashore to be found by beach strollers.

10

CARBONATE BEACHES: SEASHELLS AND THE STORIES THEY TELL

Beaches and seashells are as inseparable as bread and butter or bricks and mortar. Seashells always fascinate, both because of their amazing beauty and because of the wondrous role they play in the life of the sea. Some less beautiful, like the oysters, are delicacies that have sustained humans for thousands of years. Seashells are also important components of the ecosystem of beaches, and the remnants of their skeletons often make up a large part of the sand that is moved about in nature's most dynamic environment.

Almost all seashells are made up of calcium carbonate ($CaCO_3$) that has been extracted from seawater. There are two mineral forms with the same chemical composition but slightly different crystal arrangements: aragonite, which most snails precipitate, and calcite, produced by most clams. Shells made of calcium carbonate are referred to as *calcareous*, and the portion of the beach sediment that is calcareous is referred to as the *carbonate fraction*. Beaches that are made up mostly of calcareous materials are called carbonate beaches.

A great variety of organisms make up the carbonate fraction of beaches, and some beaches have a larger variety of species than others. For example, the beachcomber strolling along a tropical beach is liable to find a much greater number of species than would be found on temperate or Arctic beaches. Beaches that are particularly active and are affected by frequent storms usually have fewer species than those in calmer climates.

A beautiful white carbonate beach in the Maldives, the classic beach of everyone's dreams on a wintry snowy day in the northern latitudes.

Often the carbonate fraction is responsible for beach color. The pink shade of some Bermuda and Bahamas beaches is caused by tiny pink foraminifera (protozoans) that are washed ashore. The brown color of beaches is caused by postmortem staining of shells by an iron oxide (discussed later in this chapter), as seen along the shorelines of U.S. East Coast and Gulf Coast barrier islands and some east and north African coasts.

CARBONATE SHELLS, SKELETONS, AND SECRETIONS

The carbonate fraction of beaches is usually made up of some combination of the following invertebrate animal shells along with the remains of calcareous algae: mollusks, barnacles, arthropods, echinoderms, brachiopods, serpulid worms, and the microfauna that live within the beach, often between grains.

MOLLUSKS

What most people think of when the subject of seashells comes up are mollusks, of which there are four main types likely to be found on beaches. These are clams and mussels (bivalves or pelecypods), snails (gastropods), cuttlefish (cephalopods), and the less common tusk shells (scaphopods). Much of the calcium carbonate that is deposited on the world's continental shelves is from mollusks. Most shells on the beach did not live there but were washed ashore from the continental shelf or washed off nearby rocks.

Clams range in size from the giant trident clams, sometimes weighing more than 400 lb (180 kg) and found near coral reefs in the tropics, to micromollusks that require a magnifying glass to observe.

Cuttlebone, the white, oval, flat, and porous internal skeleton of cuttlefish, is a very common constituent of beaches ranging from the tropics in Southeast Asia to South Australia to the North African Mediterranean beaches. Cuttlebones are usually 2 to 4 inches (5 to 10 cm) long and were once ground up for use as a component of toothpaste. They are commonly seen in parakeet cages, where their function is to provide entertainment for the birds and calcium for the bird's diet.

Perhaps the most beautiful and sought-after shells are any of the six species of the genus *Nautilus*. These open-ocean cephalopods are rare on beaches. Among the four authors of this book, only one of us (Pilkey) has ever found a Nautilus shell on a beach, and that was on the west coast of Taiwan.

Widespread clams in beaches and sediments below the low-tide line include the quahog (*Mercenaria* sp., formerly *Venus*), razor and jackknife clams (genera of the family Solenidae), geoduck clams (*Panopea* sp.), and giant clams (*Tridacna* sp.). Other common edible clams include mussels (*Mya* sp.), oysters (species of genera including *Crassostrea* and *Ostrea*), and scallops (*Pecten* sp.). Snails include several general categories: conchs (*Strombus* is the true conch genus, but other genera go by the same name), whelks (*Busycon* and *Cabestama* spp.), cones (*Conus* sp.), cowries (*Cypraea* sp.), periwinkles (*Vinca* and *Catharanthus* spp.), abalones (*Haliotis* sp.), and limpets (*Scurria* sp.). A side note: Any snail shell that seems to be moving suspiciously fast may be occupied by a hermit crab that is "wearing" an empty shell as its mobile home.

BARNACLES

Barnacles are a type of arthropod (as are crabs) and include genera belonging to the subgroup Crustacea. Barnacles are encrusting organisms, like mussels, very commonly found on the beach (dead) attached to pieces of wood, rope, bottles, or other objects that floated ashore. These organisms secrete shells of calcium carbonate consisting of six plates connected by organic matter. The plates become disarticulated after death and are difficult to distinguish from broken-up mollusks in the beach. Such fragments are usually assumed to be mollusks. As a consequence, barnacle fragments are probably much more common in beach sands than usually recognized.

One unusual type, called the gooseneck barnacle (*Lepas* sp.), is found worldwide on beaches, hanging from driftwood. They attach to their substrate by a thick muscle (the

A beach in Laguna Madre, Mexico, made up almost entirely of small clamshells. In part, this shell concentration may have been caused by past commercial exploitation of this particular clam (i.e., this may be a shell dump from past commercial operations, although this could not be verified).

An unusual beach on the Salton Sea, California, that consists almost entirely of whole barnacles and barnacle fragments (white, purple, and pink). Some thin, elongate fish bones also are apparent. Although this example is from a saline lake, barnacle fragments are a common component of carbonate beaches everywhere. However, because barnacle plates tend to disaggregate on beaches, they are difficult to distinguish from mollusk fragments. The U.S. penny is shown for scale.

neck) that is considered a culinary delicacy in Spain and Portugal. In medieval times, it was thought that these barnacles became geese—the flying variety—hence the name.

ECHINODERMS

Spiny sea urchins and sand dollars (Echinoidea) are common components of beaches at all latitudes, but because they are quite delicate compared to snails and clams, they are usually broken up in the surf zone and, like barnacles, are not readily recognizable in the carbonate fraction of beaches. Their delicate nature is due to the fact that their tests, as their calcareous skeletons are called, consist of plates that are held together by degradable organic matter. The plates are made of magnesium-rich calcium carbonate, known as *high-magnesium calcite*. The occasional sand dollar is found whole on beaches and is much sought after by beachcombers.

Another widely distributed echinoderm is the starfish (class Asteroidea), which can occasionally be found on beaches. These too disarticulate into individual plates when buffeted about in the surf but are nonetheless more robust and more likely to be found whole than are sand dollars and sea urchins. In any case, make sure your starfish or sand dollar is long dead before you put it into your shell-collecting bag. The smell will tell you in the next day or two if you've misjudged.

CORALS AND BRYOZOA

Tropical beaches behind offshore reefs may receive a significant amount of carbonate sediment from the reef. Broken fragments of various corals (a group of reef-forming organisms within the phylum Cnidaria, formerly Coelenterata) are common, and large chunks of coral may be found on such beaches after storms. More delicate calcareous skeletal remains of lacelike Bryozoa also are common to tropical beaches but are often broken up beyond recognition.

OTHER CARBONATE-PRODUCING ORGANISMS
BRACHIOPODS

Hundreds of millions of years ago, in the Paleozoic era, brachiopods were the most common

British Virgin Island conch shells in the hands of coauthor Andrew Cooper. The slit in the top of the conch in Andrew's left hand was made by a local fisherman in order to remove the meat from the shell.

shelled animals in the shallow seas. At first glance they resemble mollusks, which have replaced them in the modern environment as a dominant marine invertebrate. The two valves of brachiopods have different shapes, as opposed to clams, which usually have symmetrical valves (i.e., valves that are mirror images of each other). The brachiopod shells found on beaches are likely to be small, less than 1 in (2.5 cm) in length. Like mollusk shells, for which they are frequently mistaken, brachiopod shells are found on beaches ranging from the tropics to the Arctic. One relatively rare brachiopod (*Lingula reevii*) was common in Kaneohe Bay, Hawaii, but ironically began disappearing when sewage was diverted away from the bay.

of dark-colored rock that is probably from volcanic bedrock.

Lower left This close-up of a gravel beach in the British Virgin Islands reveals an interesting mix of pebbles, including abundant, very well-rounded coral fragments (white) as well as fragments of coralline algae, mollusk shells, and rounded noncarbonate pebbles (dark colored). The rounding indicates that these pebbles have been on the beach for some time and have been subjected to abrasion. The credit card is shown for scale.

Upper Coral Beach, Maldives, shows a size-sorted carbonate beach with an abundance of stick-shaped coral fragments concentrated at the last storm line. The rest of the beach is fine carbonate sand, except for small cobbles

Lower right Starfish on a rippled beach. The starfish was flipped over, revealing the impression it left in the beach sand. Starfish have relatively delicate skeletons and are often broken up in the surf zone.

FORAMINIFERA

Tiny foraminifera, calcareous protozoans, seldom live in beaches but are commonly washed ashore from nearby ocean waters. These are one of the more common components of the microscopic carbonate fraction of beaches, especially along generally rocky beaches in Western Ireland.

SERPULID WORM TUBES

These marine polychaete worms secrete strawlike calcium carbonate tubes that form intertwining masses known as worm reefs. They attach themselves to rocks or large shells and often, after storms, wind up in the beach as pebbles or larger fragments.

MEIOFAUNA

This very large and loosely defined group of microscopic organisms lives within the beach, usually between the sand grains, and are mostly noncalcareous (e.g., copepods, nematodes, bacteria, polychaetes; see Chapter 9).

CALCAREOUS ALGAE

The plant kingdom also is a major contributor to the carbonate fraction of beaches in tropical areas, primarily in the form of calcareous algae. Plants are often thought of as lacking hard parts, but the humble single-celled algae often have hard parts as a component of their colony. In particular, the marine green algae genera *Halimeda*, *Penicillus* (shaving-brush algae), and *Acetabularia* (mermaid's wineglass) and the coralline algae (red algae) are important calcium-carbonate producers. *Halimeda* grows as branching colonies, with individual segments composed of the mineral aragonite, while the *Penicillus* "bristles" of the shaving brush are composed of the same mineral. When the organic matter deteriorates, tiny needle-shaped sand grains result. The internal, spherical hard part of the *Acetabularia* at its joint between the stemlike structure and the cup is also aragonite.

When these plants die or are ripped up by storm waves, the individual hard parts separate into sand-size aragonite grains that may be washed up onto beaches. With time, these particles break down into finer-than-sand-size sediment that is usually washed away from the beach. The breakup of algae and other calcareous fragments, which are much softer than quartz grains, is why the surf zone of carbonate beaches, containing carbonate mud in suspension, is often white or milky in appearance. Cancún Beach, Mexico, and various Red Sea beaches are typical examples.

The coralline algae (red algae) commonly associated with coral reefs are composed of calcite and are more resistant than other algae to breakdown into sizes smaller than

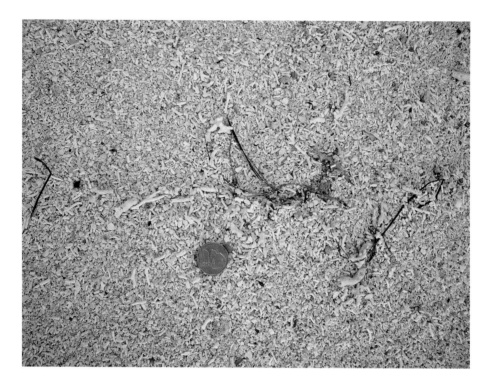

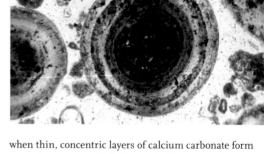

Upper Close-up of a British Virgin Islands beach reveals that it is made up almost entirely of coralline algae fragments, derived from offshore. Although the particles have been rounded slightly, this beach is subjected to smaller waves than the beach shown in the Maldives photo (page 190). The U.S. nickel is shown for scale.

Lower left Microscopic view of a Bahamian fine sand composed almost entirely of ooids, showing their typical spherical shape and polished appearance. Ooids form when thin, concentric layers of calcium carbonate form around a seed crystal, such as a shell fragment, by precipitation out of seawater in warm, shallow, wave-agitated environments. The line scale is 1mm (1,000 microns).

Lower right Microscopic view of a thin section of ooids showing the layering formed as the grains were precipitated, layer by layer. Photo courtesy of Ronald Perkins.

sand grains. Island and mainland beaches behind reefs usually have carbonate beaches that are often dominated by algal sands.

OTHER CARBONATE SEDIMENTS

Other types of carbonate sediment, including nonskeletal carbonate remains and inorganically produced carbonate grains, are important as beach materials. Aragonite mud can also form on beaches when coarser grains of aragonitic material break down into mud-size material. The sand-size grains include pellets and ooids.

Pellets are sand-size bits of animal fecal material and may be calcareous if the animal is ingesting carbonate mud or the carbonate skeletal material of other calcareous organisms. Usually pellets will be reduced to mud-size sediment by wave activity, but they are known to persist in some low-wave-energy environments and are reported from some carbonate beaches. Over time some of these pellets become cemented and hard, resistant to breakdown into mud.

Ooids, sometimes called *oolites*, are hard, spherical, calcareous sand grains that are inorganic in origin. These concentrically layered (like the cross section of an onion), spherical grains form when calcium carbonate is precipitated from seawater around a small seed grain, such as a tiny shell fragment, and the grains are rolled back and forth by wave or tidal current activity in tropical waters. The spheres resemble fish roe and are carried onto beaches from their nearby points of origin, typically warm, shallow, agitated waters of tropical shoals such as in the Bahama Banks, the Red Sea, and the Persian Gulf and along some Caribbean shores.

NONCALCAREOUS PLANT AND ANIMAL REMAINS

Several groups of animals were noted in the previous chapter in terms of the tracks, trails, and traces that they leave on beaches. Some of these animals are also important in producing sedimentary particles that make up the biogenic (nonterrigenous) portion of the beach (see chapter 3).

ARTHROPODS

Crab and lobster carapaces (hard body parts) are frequently found on beaches. Often these carapaces are molts shed by a still-living organism. These are not calcium carbonate but rather are made of an organic material known as chitin. Horseshoe crabs (*Limulus* sp.), relatives of the long-extinct trilobites, are the most spectacular crablike skeletons on beaches. As noted in chapter 9, horseshoe "crabs" are arachnids (not true crabs), a class of arthropods, up to 2 ft (more than 0.5 m) in length.

This beach on West Caicos Island, Bahamas, is made entirely of ooids. The grains form offshore and are transported to the beach by wave action. Photo courtesy of Ronald Perkins.

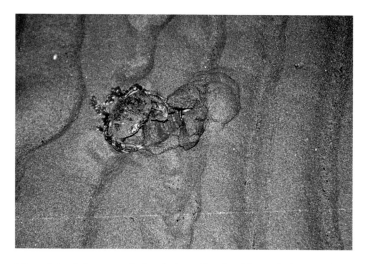

Life-and-death drama on the beach. A small crab (with attached green algae) is attempting to excavate a moon snail for its dinner on this New Brunswick, Canada, beach. This event is taking place in the intertidal zone, where ripple marks are present, and the mound in the center of the photo is an expression of the moon snail's burrow.

OTHER ANIMAL REMAINS

A few other noncalcareous organisms are found on beaches, particularly after storms, including sponges, sea cucumbers, and jellyfish. Look along the wrack line, and particularly in patches of seaweed, for these soft-bodied remains.

PLANT REMAINS AND DRIFT SEEDS

Seaweed can be the most abundant organic component on many of the world's beaches; it is particularly abundant in the wrack line. Sometimes seaweed layers are buried in the beach and become important to the beach's food chain and in forming a chemical-reducing, low-oxygen-content environment that may alter sediment properties such as shell color (discussed later in this chapter).

Drift seeds are another plant component found on beaches. These long-floating seeds, typically from tropical plants, have been used by oceanographers to chart ocean currents. Some drift seeds can float for a couple of years and travel very long distances. Coconuts from tropical sources, for example, have been found as far north as Norway.

Some important drift seeds and their points of origin include coconuts and coral beans from the tropics, anchovy pear, sea heart, and crabwood from the New World tropics, box fruit from Polynesia, the sea bean from the Indian Ocean tropics, coco-de-mer from the Seychelles, and puzzle fruit from Southeast Asia. Among some beach aficionados, drift seeds are as important as shells.

SEA GRASSES

Although not direct producers of sediment, sea grasses play an important role in trapping carbonate sands and muds in the nearshore zones of tropical beaches. Meadows of sea grass grow in shallow water on the submerged part of the beach, often extending right up to the low-tide line. These sea grasses bind the sand grains together and impart cohesiveness to the seabed. In some instances, the grasses can even form a kind of turf on the seabed. During storms, sea grasses are washed ashore and often accumulate in thick deposits at the wrack line, even covering the beach (see Chapter 3).

Seaweed that has accumulated along the shoreline at Ballywalter, Northern Ireland, obviously makes beach access, strolling, and other forms of recreation difficult.

SHELL ABUNDANCE

As anyone who has strolled on a lot of beaches knows, the shell content of beach sand can be highly variable. Some natural beaches are so shelly they are impossible to walk on with bare feet. Others are so "unshelly" that one has to search diligently for the occasional shell. Beach shells come from many sources.

There are some instances where beaches on adjacent barrier islands, in seemingly similar environmental conditions, exhibit large differences in their shelliness. Such is the case for the beaches of Shackleford Banks and adjacent Bogue Banks, North Carolina. Shackleford has the region's most shelly beach, while Bogue's beaches contained few shells (prior to an artificial dredge-and-fill beach-nourishment project that brought in an oyster-shell hash).

There are two principal reasons why the shell content (the carbonate fraction) of beaches can vary so much: *productivity* and *dilution*.

Shell productivity, the variety of species and abundance of individuals, is determined by water temperature, the most important control on shell abundance globally. All other things being equal, the warmer the water, the higher the beach's shell content. Calcareous marine organisms are more abundant and more productive in warm water, which explains the frequent occurrence of pure carbonate beaches in the tropics and in warm waters such as on atolls, the beaches of the Caribbean and the Red Sea, and on some Southeast Asian shores. Thus, beaches in Morocco are dependably more shelly than beaches in Norway, as are beaches in Florida compared to those in Labrador.

A large accumulation of mangrove seeds at the high-tide line on one of the Brazilian Gurupi Islands south of the Amazon River. These seeds from mangrove forests along estuarine shorelines floated out to sea through inlets. Mangrove seeds are sometimes transported hundreds of miles before being deposited on a beach.

The most prolific calcium carbonate producers are coral reefs. It is common for beaches that are sheltered by coral reefs (e.g., in the Caribbean and the Red Sea) to consist of pure calcium carbonate and contain abundant fragments of corals, shells, and calcareous algae. But there are many exceptions to this rule of water temperature and shell productivity. Purely calcareous beaches exist in some far-northern areas (including Labrador, Norway, and Southeast Alaska). These occur where the rocky seafloor or nearby cliffs are covered with attached mussels, barnacles, sea urchins, algae, and foraminifera, which are torn off by storm waves and piled on the beach. Such beaches are free of abundant pebbles or boulders; if these hard rock components are present, they will crush and grind the much softer calcareous fraction to powder in the erosive surf.

An example of this effect is Sand Beach, Mount Desert, Maine, a pocket beach between granite headlands. Storms tear the mussels, barnacles, limpets, and sea urchins off the nearby rocks and toss them onto the beach, where wave action quickly grinds the shell debris into sand-size material. In this fashion, a calcareous beach is born in cold water.

Dilution by nonshelly sediment, usually at river mouths, is another one of the controls on carbonate content in beach sand. In the Virgin Islands and along the north shore of Puerto Rico, for example, there are glittering white and light-brown pure-shell beaches only a short distance from blackish gray–sand beaches, derived from volcanic bedrock, that contain few seashells. The dark-sand beaches on these islands are found at river or stream mouths, where fresh sand from the interior is provided to the nearby beaches during every flood. This noncarbonate fraction dilutes the otherwise high carbonate background content.

Collection of tropical seeds found on the beach of one of the Gurupi barrier islands in Brazil. To the right is a mangrove seed that already has a sprouting leaf. The toe of the shoe is shown for scale. Photo courtesy of Allen Archer.

A sea-grass bed exposed at low tide on a carbonate beach on Great Camanoe, British Virgin Islands. Sea grass causes sediment to accumulate and also holds sediment in place, offering resistance to sand transport by waves. In the background, multiple wrack lines are visible at the back of the beach, marking storm-tide or high spring-tide levels. Most of the wrack material is sea grass. Note also the well-developed swash marks and ripple marks.

This carbonate beach in Hamerfest, Norway, above the Arctic Circle at 71 degrees north latitude, is an exception to the rule. Carbonate beaches are found mostly at low latitudes in warm water, but in the absence of noncarbonate sand in high latitudes, and in the presence of shelled organisms to produce sediment, these unusual beaches can form. The shells on this beach were derived from offshore and from nearby rocks, from which storm waves tore them away.

On a larger scale, the beaches on the west coasts of North and South America tend to be less shelly than those of the east coasts of these continents. This difference is because of the close proximity of west coast mountain ranges, from which fast-moving streams carry much sand directly to beaches; this sand contains no shell material. Even at the equator, where carbonate productivity is high, the Pacific Coast beaches of Colombia and northern Ecuador contain relatively few seashells. The nearby Andes Mountains furnish massive amounts of sand through numerous streams, large and small, spurred on by the locally high rainfall of the tropical rain forests.

On the Atlantic side of North and South America, mountain ranges are well inland from the ocean, and streams tend to dump their sand loads at the upper ends of estuaries, far from the beaches. These estuaries are river valleys drowned by the last sea-level rise. Thus, dilution of the carbonate fraction of beach sand by river sand is relatively unimportant because the river sand does not make it to the beach. Exceptions are the beaches near the mouths of large rivers such as the Mississippi, Magdalena, and Oronoco.

WHERE DO SEASHELLS COME FROM?

On most shores, the shells we hunt for on the beach are derived from organisms that lived off-shore, near the beach; after they died, the shells were carried ashore by waves and currents. Some organisms live in the beach itself, and you may find the occasional snail or happy clam, which is best left alone unless it is for the stew. For the most part, however, the shells we collect no longer have an organism living inside. Years or decades may have passed while the shell journeyed to where it is found. For some beaches, a shell's history is even longer and more complex. This is especially true on barrier islands and trangressive coasts, where the rising sea is rolling over earlier marine sediments and reworking the contents, including buried shells. Dating of some shells on southeastern U.S. beaches has given ages in thousands of years, old enough to be considered fossils, even though the shell may appear as if it were lived in yesterday!

A shell lag on a beach near Kashima, Japan. This same accumulation of shells on a southern U.S. beach would be extensively brown stained, but colder waters such as those off Japan do not favor brown staining. Note the predominant concave-downward orientation of the shells on this beach. As explained in the text, this is the hydrodynamically stable orientation of shells on open-ocean beaches. Unfortunately, a plastic bottle is visible for scale.

By definition, a fossil is some remains of an animal that lived in the geologic or prehistoric past. These shells on barrier island beaches that are hundreds and even thousands of years old, in part, tell the story of the immediate geologic past. The organisms lived at a time when the level of the sea and the temperature of the ocean water were slightly different from what they are at present. Now that beach nourishment is so common for tourist beaches, some of the shells found on such beaches may have been dredged from sediments that are a few million years old.

Seashells on a particular beach may come from any of the following sources:

· *Animals that lived within the beach* are an important source of shells on some beaches but not on others. The surf zone is a very harsh environment for living things. Besides the pounding waves, there are also wide ranges of temperatures and alternate wetting and drying as tides move up and down. Typical of harsh environments, the numbers of shell species that live within the beach tend to be small, but the number of individuals is often large. Purchase of a local shell book will provide information about which shelled

An accumulation of razor clam shells from the Sefton Coast of England. The black objects among the shells are chunks of peat derived from nearby peat outcrops on the beach. A shell accumulation consisting mostly of a single species is caused either by the animals' great abundance near the beach or by selective physical sorting of the shells in the swash zone based on their shape and buoyancy (or by humans, in the case of middens).

animals actually live within your beach. One of the most familiar is the coquina clam (*Donax* sp.), a thumb-size edible clam (also known as cockles and pipi). There are nineteen *Donax* species around the world, and they live on just about every beach in the world where people are likely to swim.

· *Shells from animals that lived offshore* are driven ashore by the waves. A wide variety of clams, snails, sand dollars, and sea urchins, discussed earlier in this chapter, live on the inner continental shelf. Similarly, reefs and meadows of sea grasses are offshore environments that produce an abundance of shelly fauna that are likely to be washed up onto adjacent beaches. Those who visit beaches at different times of the year can observe that different species of shells come ashore with the changing seasons, in part because of varying wave conditions. Pen shells (*Pinna* sp.), of which there are 250 related species around the world, are large fan-shaped clams that partially bury themselves in the sand on the seafloor. During storms they are readily torn away and wind up on beaches, especially in the winter.

· *Shells that have been peeled off nearby pilings and rocks by incoming storm waves* can be a major component of pocket beaches on rocky coasts. Common constituents of a beach fauna derived from rocks would include barnacles, limpets, mussels, and other shells (as discussed earlier in this chapter). On sandy coasts without naturally occurring rocks, such shells may be contributed to the beach from rock seawalls and jetties and even from pilings on nearby piers and docks.

· *Fossil shells derived from ancient rocks that outcrop on the beach or offshore* can be very important locally. Many continental shelves are not completely covered with sand, especially those that are far from any major river mouths, and old rock outcrops are frequently exposed to seafloor weathering and wave activity. The fossils that are pre–Ice Age (3 million years or older) generally

A beautiful white-sand beach with abundant coral fragments in Kenya that is protected by an offshore coral reef marked by the white line of breaking waves. The sand is white because most of it is calcareous material derived from the coral reef and its associated organisms.

A view of misnamed Coral Beach at Galway, Ireland; there are no coral fragments in the sand and no corals offshore in the cold water off Ireland. However, the beach is made up of an unusual concentration of almost pure calcareous sediment (coralline algae and mollusk shells), hence its name.

are very old in appearance and often are from species that no longer exist. An example of this shell source is the occurrence of 20-million-year-old shells of *Ostrea gigantissima*, an oyster that is almost 2 ft (60 cm) long, on North Topsail Beach, North Carolina.

· *Lagoon shells in lagoonal sediments that were overrun by a migrating barrier island* are a source of relatively young (a few thousand years old) fossils. These old lagoonal deposits are a particularly important shell source on barrier island beaches. Barrier islands can migrate in a landward direction in response to a sea-level rise. When this happens, the islands overrun mud from mangroves or marshes and sand deposits from the bay or lagoon originally behind the island. This migration is evident from the mangrove or salt-marsh mud deposits on the open-ocean beach in places as varied as Ossabaw Island, in the U.S. state of Georgia; Santa Barbara Island, Colombia; and Xefina Island, Mozambique. Very common examples of this type of shell on barrier island beaches are various species of oysters that grow exclusively in quiet, lower-salinity bay waters and not in the open ocean. Such shells on beaches can be almost any age, but most are a few thousand or tens of thousands of years old, not counting those left behind from oyster roasts and clambakes.

· *Shells from artificial beach fill* are commonly fossils, hundreds to thousands of years old and from any environment depending on where the truck or dredge has obtained sand to construct the artificial beach. Beach sand in Waikiki, Hawaii, was once brought in by freighter from California and Australia. Shelly beach sand with Gulf of Mexico shells was once brought from a Gulf of Mexico beach across the Florida peninsula to Miami Beach, where a hotel owner wanted to improve the shell hunting! Ordinarily, a seashell species assemblage on a beach that is very different from nearby local beaches (that are still in a natural state) can readily identify an artificial beach. For example, some artificial beaches are derived from dredging lagoon sediments and thus have abundant lagoon shells, such as oysters, on the open-ocean beach, where they do not belong. In addition, shell coloration may provide a clue as to the origin of the shell. If most shells are black or gray, the beach may be artificial. Shell color is discussed later in this chapter.

THE SIGNIFICANCE OF BROKEN SEASHELLS

Shells on beaches can be broken up for a number of reasons, some natural and some due to human activities. The relative importance of the two will vary widely from beach to beach. In the human-activity category, driving vehicles on the beach and dredging are the usual suspects, whereas shell-cracking animals and crashing waves are natural causes of shell breakage.

Shells display a wide range of vulnerability to breaking. Animals that live in environments exposed to large waves tend to have robust shells that may survive on a beach surface. Animals that live in more sheltered habitats, such as sea urchins, sand dollars, and other echinoids, are usually far more delicate (as mentioned previously) than most clams and snails, which is why they are not often found whole on a beach.

Shell-breaking processes include the following:

Vehicular traffic on beaches causes shell breakage and sediment compaction and impacts burrowing and nesting animals. This example from Portstewart, Northern Ireland, illustrates the conflicting problems arising from multiple beach uses.

- *Beach driving.* Obviously the continuous back-and-forth movement of wheeled vehicles can break up the more delicate shells and sometimes the hardy shells as well. Many of the world's developed beaches allow some sort of vehicular traffic for sightseers and fishers—even in national parks. Some beaches, like Daytona Beach, Florida, were once the site of car races. An exception is South Africa, where in 2006 virtually all driving on beaches in the entire country was prohibited. Unfortunately, on some U.S. National Seashores' wilderness areas where driving is otherwise prohibited, the National Park Service insists on sending a daily four-wheeler patrol up and down the beach—crunching shells all the way and leaving unsightly tire tracks. Wilderness areas need not be patrolled in this fashion. On many developed beaches, the daily or weekly beach cleanup consists of some sort of mechanical rake that is dragged along behind a tractor, which can readily break up shells.

- *Dredges.* Sometimes a cluster of shell hunters can be seen gathered in a semicircle around the end of a pipe that is spewing a slurry of offshore sand and water onto a beach. Every once in a while someone darts in and triumphantly grabs a shell. In some dredging projects, however, few shells survive the long path from the offshore dredge through the pipe to the beach, and most shells come ashore in fragments. One can hear the clanging and banging of shells as they move along the pipe. Probably the degree of breakage depends on the distance piped and the concentration of the slurry as it rushes through the pipe. We observed one nourished beach project on Emerald Isle, North Carolina, where even the thick shells of the up to 4 in (10 cm)-long *hard clam* or *Northern Quahog* "Mercenaria" clams were fragmented. On a beach where almost all the shells are broken, the usual suspect should be dredging.

Upper left This calcareous beach on Barbuda in the Caribbean shows the famous pink-to-red color that is attributable to a form of calcareous algae that washes up onto the beach. The sand is highly fragmental but is slightly more rounded than the carbonate beach sand shown from Galway, Ireland.

Lower left Black-stained shells on a nourished beach on Bogue Banks, North Carolina. In most beaches, an accumulation of pure black shells is almost certainly evidence of a nourished beach. Black sand comes from an environment that is without oxygen, and often that environment is at a burial depth of 3 ft (1 m) or more in continental shelf sediment. In some communities, engineers have gone to a lot of extra expense to find sand that is the same color as the original beach sand.

Upper right This beach shell lag has shells that are highly fragmented, and many of them are brown stained and show no evidence of abrasion from wave activity. The small clamshells are mostly "correctly" oriented, with their concave sides down.

Lower right This close-up of a shell lag on Core Banks, North Carolina, shows the highly rounded shells that are typical of high-wave-energy beaches. The reworking and abrasion of the shells gives some of them a natural polish, and although the calcium carbonate is soft, these rounded beach shell fragments are sometimes used to make jewelry. Photo courtesy of Rob Greenberg.

- *Predators.* If a lot of fragmented shells are found on beaches with generally small waves and no driving, it is a sure indication that something else is responsible for the breakage. That something could be a wide range of animals, vertebrate and invertebrate predators, that are busy every day cracking shells to get at the meat inside. A variety of crabs and rays cruise the seafloor grabbing any shelled animal unwary enough to come out of its burrow at the wrong moment. Crabs, especially those with one oversized claw such as the Florida stone crab (*Menippe* sp.) and the European green crab (*Carcinus* sp.), specialize in attacking snails. Rays can crack large, thick shells between two hard plates in their jaws. Some fish species also have jaws strong enough to crack shells. Parrot fish contribute to the carbonate fraction of beach sediment where coral reefs are offshore. These fish scrape algae off coral heads with their parrotlike beaks and in the process create a lot of coral-fragment sand grains scraped from the coral heads. Then there are the seagulls that have learned to drop shells on concrete sidewalks and pavements to break them open. When gulls drop shells on rocky beaches to break them, the fragments will become part of the beach sediment.

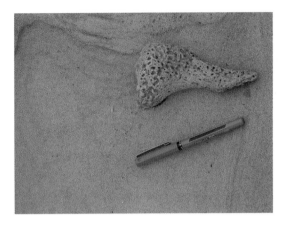

This highly degraded conch shell from the beach on Ocracoke Island, North Carolina, is likely to be many years old, as evidenced by the numerous boreholes in its surface. A variety of organisms, including a type of sponge, bore into shells to obtain nutrition from the organic matter, or into empty shells to seek shelter. This boring contributes to the weakening of the shell, and it will someday be reduced to sand-size fragments.

· *Waves.* Breaking storm waves can resuspend beach sediment, causing a mixture of sand and shells to churn violently about in the turbulent surf zone. On rocky coasts, the shells torn off the cliff by storm waves and washed ashore to the nearby beach sometimes crash into rocks and break up along the way.

SHELL ROUNDING

As shells break and tumble about in the surf zone, their corners and edges slowly become rounded and less sharp. The more wave energy there is and the longer a shell is on a beach (and not buried or brought onto land by storm overwash or carried offshore by a storm), the rounder the shell becomes.

Sometimes the rounded shell fragments display beautiful patterns of color that are used for jewelry. Unfortunately, the relative softness of shell material makes the jewelry a bit delicate. Beautiful rounded fragments of shell can be found on just about any sandy beach that has frequent high waves. Examples include the Outer Banks beaches near Cape Hatteras, North Carolina, beaches on the north side of Puerto Rico, and the beaches of western Morocco and northern Tunisia.

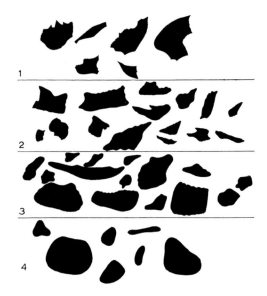

This roundness scale for the carbonate fraction can be used to compare shell fragments on a beach to the chart silhouettes to quantify their stage of rounding. Box 1 represents highly angular fragments at initial breakage. As abrasion occurs, shell fragments become subangular (box 2) to subrounded (box 3). For the most part, highly rounded shell fragments (box 4) are found on beaches with high waves. Beaches with vehicular traffic may have artificially induced poor rounding of shell fragments by breakage that increases angularity.

ORIENTATION OF SHELLS ON THE BEACH

Shells on beaches often exhibit preferred orientations; the preferred orientation refers to the position attained on a beach by the majority of the shells. The shell position or orientation can refer to the alignment of a shell relative to the shoreline, the waves, or the wind direction. On a beach with significant wave activity, shells will orient in the most stable position—that is, a position most likely not to change with time, a position that the shell maintains because it offers the most resistance to the forces of waves and wind. The long axis of elongated shells (those that are longer than they are wide), such as oysters and razor clams, can clearly show a preferred orientation. Snails and larger shells such as conchs have a narrow end and a wide end, which can be the basis of a preferred orientation.

Sometimes wind creates the preferred orientation. In strong winds, small shells may be picked up and redeposited. Also in strong winds, large conchs and whelks may be turned to a preferred orientation as the wind excavates sand along the shell margin, slowly causing the shell to move. It is likely that under the right circumstances, wind and water may act in concert to orient shells.

Clamshells lying on a beach can show a preferred orientation with respect to which side is up. A clamshell has a concave side, where the animal once lived, and a convex or curved outward surface, which faced the outside world when the animal was alive. Such a shell lying on a beach can be oriented one of two ways: concave side up or concave side down. On most beaches, there is a strong preferred concave-down orientation of clamshells. Typically 80 to 90 percent of such shells will lie on the beach with the cavity that once protected the living animal facing down. The reason for this orientation can be quickly demonstrated by placing a clamshell in the concave-up orientation within the swash zone. The shell will be quickly turned over by the swash and reoriented with the concave side down. Subsequent swashes may move the shell about but usually will not turn it over. The concave-down orientation is known as the hydrodynamically stable orientation.

With the aid of a mask and snorkel, one can wade out and watch the process occurring in hip-deep water (providing the surf is not too large) but still within the surf zone. Releasing a clamshell at the surface of the water column reveals that the shell will immediately attain a concave-up orientation as it rocks back and forth on its way through

Roundness of sand-size material also reflects the degree of abrasion to which the sediment has been subjected, as in this carbonate beach sand from the island of St. Maarten in the Caribbean. Carbonate sand grains often show glossy, polished surfaces, as in this example. Several of these grains are foraminifera (e.g., lower right margin). The line scale is 1mm (1,000 microns).

A brown-stained scallop shell is residing in a stable concave down position. Notice the current scour around the shell.

the water column to the seafloor. Concave up is the hydrodynamically stable orientation for falling shells. They land on the bottom still concave up. But if the surf is strong enough, the shells will be quickly turned over to their more stable concave-down orientation.

Now go farther offshore to the realm of scuba gear, to a depth of 30 ft (9 m) or more. A shell released in the water column here will settle down concave up, just as it would in the surf zone. But once the shell reaches the seafloor, it remains concave up because the waves aren't generally strong enough to turn over the shell. In fact, early studies of bottom photographs of the surfaces of continental shelves revealed that most shells offshore are oriented con-

The clamshells on this beach adjacent to a jetty at Oregon Inlet, North Carolina, are all oriented "correctly" in the concave-down position. The shells seen here on this open-ocean beach are primarily lagoonal species, not normal open-ocean forms, suggesting that they were dredged from an offshore exposure of old lagoon deposits that were left behind as the barrier island moved landward in response to rising sea level. In addition, note that most of the shells are white and a few are black, but there are very few brown shells. The species makeup and shell colors strongly indicate that this is a nourished beach. Note also that the beach surface is pitted with raindrop impressions.

This diagram shows how the hydrodynamics of a shell determines its orientation under different circumstances. A shell sinking in quiet water often takes a concave-up orientation (depending on its shape) but may be oriented to a convex-up position when turned over by a wave or current. Drawing by Charles Pilkey.

cave side up. Even though storms can occasionally kick up the shells, they tend to resettle in concave-up orientation.

One way that geologists who study ancient sandstones can distinguish between a beach deposit and a continental shelf deposit is by observing the orientation of the fossil clamshells (if they are present). In fact, geologists' interest in interpreting the paleoenvironments in which sandstones and limestones formed is one of the reasons why we know so much about the surface features of beaches that we discuss in this book.

SECONDARY SHELL COLOR

Many of the shells on the beach are not the original color that they were when the animal that lived in them was alive. Such secondary colors are imprinted in the shell after the animal has died. Brown staining of shells, or the common light yellow-brown stain on shells, is probably caused by the formation of *limonite*, an iron oxide, in the microscopic interstices of shells. Apparently the brown coloration is given to the shell while it resides exposed to the air on or near the surface of the beach. The iron that combines with the oxygen from the atmosphere to form limonite may come from the degradation of iron-bearing heavy minerals (the black sand) in the beach sand.

Although brown staining does not occur offshore on the continental shelf off the southern United States, apparently it does on the shelf off Puerto Rico. The evidence for this is pretty overwhelming, since, down to a depth of 150 ft (more than 45 m) or so, the entire north shelf of Puerto Rico is paved with brown-stained mollusk shells, while the continental shelf off the southeastern United States has only a few patches of brown shells. Perhaps the warmer and clearer water and the more energetic wave climate in Puerto Rico are responsible for this phenomenon. Whatever the cause, it is clear that brown-stained shells are created in different ways on different coasts. It is also an illustration of an old principle of beaches: Be very careful when transferring generalizations about beach processes across state lines, and even more so across oceans!

Where brown staining occurs in widespread fashion, it is often responsible for the overall light-brown color of beaches that is particularly characteristic of southern U.S. beaches. Brown-stained shells are probably a common global occurrence on beaches in all climates except perhaps Arctic beaches, but no studies have been made to quantify this.

The brown staining seems to be durable as long as a shell resides on the beach. If it is washed onto the upland by storm overwash, the brown staining slowly disappears, and after a few years of residence on land, the shell is bleached to gray. Bleached shells are common on overwash fans and also along the margins of flower gardens and on outdoor railings where avid shell collectors have placed them. With time, usually within two or three years, all shell coloration, original or secondary, disappears when the shell is exposed to the elements without occasional inundation by seawater.

Black-stained shells are at least as abundant and as widespread as brown shells, perhaps more so. The black staining forms in an entirely different fashion from brown staining. The black color is probably due to microscopic crystals of iron sulfide that formed within the microscopic interstices of shells, and the crystals form in the absence of free oxygen. Burial in mud or deeply in sand 3 to 6 ft (1 to 2 m) provides the requisite oxygen-free environment, so instead of oxides forming, as in the case of the brown shells exposed to the air, sulfides form.

One can easily demonstrate this process by taking a few fresh shells and burying them in mud, preferably salt-marsh or mangrove mud. Within a few weeks, many of the shells will be blackened. Some shells blacken within days of burial, and others, such as the scallop shell, don't blacken after months of burial.

Mud is not the only environment in which shells will blacken. On many beaches around the world, seaweed is buried in the sand, and as the plant material rots it forms a microenvironment devoid of oxygen. Shells readily blacken when buried in such layers under these circumstances. We have seen this type of blackening on the beaches of Portugal, South Australia, Brazil, Tunisia, northern France, and western Florida, among other places.

In the absence of buried seaweed on a barrier island beach, the presence of black shells is good evidence of island migration. For example, along the U.S. East Coast from Long Island, New York, south, there is essentially no mud on the continental shelf. The only marine mud along this entire stretch is found in the bays behind the islands, so black shells on the beach must have once resided in the bays, where there is lots of mud to blacken shells. This being the case, the only logical way that large amounts of black shells can come to a barrier island beach is for the island to migrate over a marsh deposit. What was once on the back side of the island is now on the front side. Of course, blackened shells on artificial beaches are a reflection of the sand's source area, which may include buried marsh sediment from a lagoon or sediment out on the continental shelf left behind as the sea level rose.

probably by a combination of vehicle traffic, stingray activity (crushing), and wave action.

Center Microscopic view of brown-stained calcareous sand from the beach at Luquillo, Puerto Rico. This fine sand is nearly 100 percent broken skeletal material from nearshore organisms. Note the many small, stick-shaped grains derived from echinoid spines, as well as clamshell fragments and a foraminifera shell in the center of the photo (coiled). The grains show polish as well as the brown stain.

Lower Shells showing varying degrees of black staining, including marine mollusk shells as well as oysters that once lived in a lagoon or estuary. The black color is secondary, forming after the organism died, and is caused by very fine iron sulfide in the interstices of the shell. This compound is formed when the shells are buried where there is no oxygen, as in mud, in rotting seaweed, or in the sediment at depths of 3 to 6 ft (1 to 2 m), commonly on the adjacent continental shelf.

Upper Shell hash on an Ocracoke Island, North Carolina, beach shows shells that are predominantly brown stained, a common color for beach shells of the southeastern United States. Most of these shells have been broken up,

Buyer beware. This boatful of Caribbean conch shells is not on a Caribbean island but in Ocracoke, North Carolina. These shells, probably from the Bahamas, were for sale for nine dollars each, and this same beautiful species is found at shell shops all over the world. Killing of shelled mollusks for the tourist trade is threatening the existence of some species and has impacted subsistence-food fishing for some islanders.

The continental shelf off the eastern United States has abundant black shells in its sediment cover, a reflection of a history during the last sea-level rise, from eighteen thousand years ago to the present, of barrier islands repeatedly migrating over bays. Thus, old bay deposits with their black shells are smeared over much of the surface of the continental shelf.

It appears anecdotally that the staining in black shells is much more durable than brown staining. Black shells eventually bleach when stranded away from a beach, but at a slower rate (decades rather than the few years required to remove brown staining).

SHELL COLLECTING: AN ENVIRONMENTAL AFTERTHOUGHT

Picking up shells on the beach is a relatively harmless hobby. Who among us on a beach outing can resist collecting a souvenir shell or two for displaying on the bookshelf or desk back home? They are objects of wonder and beauty and remind us of good times and a great environment.

Is this a beach of popcorn? The answer lies on the beach at Fuerteventura in the Canary Islands, where this white, popcornlike material occurs. It is made up of fragments of calcareous algae (red algae), washed up onto the beach and bleached white. The white algal nodules are mixed with black gravel, which consists of volcanic rock fragments (basalt and some black volcanic glass) derived from the underlying wave-cut platform. The camera lens cap is shown for scale. Photo courtesy of Ignacio Alonso.

On the other hand, commercial shells in shell stores are most often collected in bulk while the animal is still alive; the animal is then killed and the shell cleaned before being offered for sale. One clue to the fact that such shells were not picked up on beaches is that, as a rule, such store-bought shells never show evidence of the abrasion or wear caused by rolling around on a beach.

The foot-long Bahamian conch, or queen conch, is an example of a widely sold shell, recognizable by its large size and bright pink color in the shell opening. Every shell store in Europe and America seems to have these beautiful shells in abundance. They are used to make trumpets or souvenir home decorations, and the meat, which is a staple for Bahamians, is even considered to be an aphrodisiac. Now the Bahamian government says the shells are in danger of disappearing in the vicinity of settlements. This story is repeated throughout the Caribbean. A Roatán (Honduras) islander told us he once could obtain the family's evening meal by thirty minutes of shellfish fishing; then a commercial shell-collector supplier overfished the grounds in less than two years, and today the man may have to fish all day to secure the family meal.

Rare species, such as the Nautilus shell, are sold individually. Less rare forms may be sold by the bucketful, ten to twenty U.S. dollars per bucket. Such exploitation of marine mollusks, echinoderms, corals, and other marine organisms has a negative impact on the ecology of collecting locales and, in the case of rare species, may impact the species' very existence. Sea turtle shells are another case in point but are rarely seen in shell markets today.

It's more fun (and more difficult and challenging) to find your own shells and treasures.

11

DIGGING THE BEACH:
INTO THE THIRD DIMENSION

Many beachgoers observe features on the surface of the beach, but small children with pails and shovels, and geologists with bigger toys, enjoy looking deeper. To understand the inner expressions of the surface structures and to piece together the sequence of recent events that formed and shaped the beach, this third dimension must be considered. Natural exposures of the beach subsurface are not abundant; however, eroded beach scarps and the vertical cut banks of stream channels through a beach provide places to look for clues of its recent history. More commonly, those who study beaches dig shallow trenches with shovels and trowels to expose the sediments. The layers or beds of sand are like the pages of a book, and the vertical cross section of a beach gives us more of the story than just the surface page.

Digging deep trenches in a beach, however, is unsafe, and we caution readers not to dig deep holes, and especially to supervise children, who may dig too deep for their safety. Instead, to reconstruct the longer history of a beach, geologists drill completely through beaches (a process called coring) or use remote sensing devices such as ground-penetrating radar or seismic units to create an image of the subsurface sedimentary layers, their structures, and the position of the water table.

BEDDING

Whenever sand is deposited by wind or water, the resulting organization, or structure, in the accumulation of grains depends on the way the sediment was deposited. For ex-

This wave-cut beach scarp on Mexico's Yucatán coast shows parallel laminae and bedding in the sediments. Such scarps and stream-cuts across beaches expose the internal bedforms in the sediment layers. Parallel laminae are most common, but various types of cross bedding are also common. Photo courtesy of Miles Hayes.

ample, a sand wave or megaripple will show inclined or sloping laminae or beds, commonly called *cross bedding*. Or, as noted in chapter 6, where the water from a breaking wave washes up the beach and back, a thin deposit of sand grains, often no thicker than the individual grains, accumulates when the water of the swash sinks into the sand. Wave after wave brings in more sand, and layer upon layer of sand grains accumulate as laminae, which are often parallel to one another. An individual lamina may be distinguished from other laminae by slight variations in grain size, mineral composition, or color. A related sequence of laminae that accumulates together in this fashion is a *plane bed*, given this name because of the parallel, planar nature of the layers.

The grain size within beds reflects the energy of the event that deposited the layer as well as the availability of the sediment. For example, with the onset of a storm, the waves become larger and carry and deposit larger grains (coarser sand, gravel, and even large shells). Initially the storm waves may erode or plane off the upper part of the beach surface. As the storm ebbs and the waves begin to lose energy, the very coarse material will be deposited first, usually on top of and parallel to the earlier, finer sand beds. As the storm continues to wane, the waves become smaller and the size of the grains that accumulate becomes smaller as well. Thus, the new bed of laminae that forms will contain sediment that becomes progressively finer from bottom to top. The resulting bed is called a *graded bed* because the size of the sediment grades from coarse at the bottom to fine at the top. This ideal situation of waning wave energy and graded bedding does not always occur, however. A similar graded appearance can result when a shell or gravel lag deposit develops and is then buried by finer sand from wave swash.

Although the beds of sand and shells that form on a beach are parallel to one another, they are rarely perfectly horizontal. Beneath the seaward-sloping swash zone, the parallel beds slope seaward, usually at less than 10 degrees, matching the sloping surface of the beach. Beneath the backshore, which receives sand only from waves at high stages of tide or during storms, the beds are closer to horizontal. On tidal flats, the cross bedding of megaripples or the beds deposited on the face of a migrating sandbar will have steeper inclinations, up to 22 degrees. Where such dipping beds are either deposited on top of horizontal beds or are buried by horizontal beds, angular bed contacts result. Tidal cycles leave distinct records in cross section, showing, for example, how plane beds may be scarped by erosion on the high tide, remain exposed on the low tide, and then become buried under a drape of new sand layers on the next high tide. There are an infinite number of combinations of erosional and depositional sequences related to tidal cycles, producing a great variety of patterns in the bedding.

BLACK SANDS AND CROSS BEDDING

On the landward edge of the beach, near where the sand dunes begin, and in the intertidal zone between the berm and the longshore bar, there is more complexity in the structure of the sediment. Near the dunes, wind processes alternate with wave processes to leave a record of the changing conditions. After a storm erodes the dunes, beds of wave-deposited sand, often rich in dark-colored heavy minerals (see chapter 3), form at the surface of the beach. These black sand beds are often several inches (centimeters) thick, depending on the duration of the storm and the associated waves that concentrated them. When calmer conditions return, dune grasses again thrive, and wind becomes the primary agent of sand deposition, laying down finer sand-size material atop the storm deposit. The windblown sand lacks thick concentrations of the dark heavy minerals, but occasional thin black laminae of heavy minerals may highlight the internal bedding.

Sand that is transported by the wind travels in ripples across the surface of the dune and then avalanches down the leeward dune slope (see chapter 8). The dune's resulting internal bedding is inclined at the angles of the sloping dune faces, giving rise to eolian cross bedding. These cross beds may be inclined by up to 30 degrees, the natural angle of repose of dry sand. In exposed natural gaps through dunes where the wind has had an erosive, etching effect, cross bedding often stands out in relief. Where dune sand has accumulated on top of beach sand, the two may be distinguished by grain size and type of bedding. If there is coarse material in the sand, such as shells and shell fragments, the wind preferentially removes the finer sand, leaving a lag deposit or coarser-grained pavement that resists additional wind removal of sand. These coarse pavements form distinct horizons when buried.

Seaward of the berm, the intertidal zone of the beach has sediment that is packaged in structures formed by waves and currents. Ripple marks, discussed in chapter 6, are one of the common surface structures that have a distinct internal pattern in the beach-trench cross section. Ripples migrate by a process not unlike that of sand dunes, but in a much smaller dimension, so the sand is deposited on the face of the migrating ripple as inclined laminae. In cross section, these ripple laminae are referred to as cross laminae to distinguish them from the thicker beds of cross bedding that are seen in wind-formed dunes and water-current-formed megaripples and sand

Upper left A cross section through a megaripple shows cross bedding in which the layers slope in the direction that the current was flowing. The edge of the machete parallels this lamination. Imagine this feature being buried by horizontal layers of beach sediments. The cross bedding would be preserved as evidence of this original megaripple. Photo courtesy of Miles Hayes.

Lower left Trench exposure from bottom to top indicates deposition by a migrating sand wave of some sort, followed by the horizontal parallel laminae of beach swash, then another episode of small megaripple formation and migration to the right, followed by a return to swash conditions. Note the variation in sand-grain size between the individual laminae. The pencil is shown for scale. Photo courtesy of Miles Hayes.

Upper right When trench faces are cut at right angles to each other, the three-dimensional aspect of the beds may be seen, and the relationship between surface bedforms and the internal structure can be determined. Here, most of the upper part of the section consists of parallel laminae deposited by wave swash. In the midface to the left of the scale, some ripples are apparent, and the U-shaped pattern to the right of the scale may be buried footprint cross sections (the weight of the foot having deformed the horizontal laminae). Photo courtesy of Miles Hayes.

Lower right A trench exposure in a Massachusetts beach shows mostly the horizontal bedding in sediments that accumulated on the surface of the beach berm. The angular contact to the left of the measuring scale resulted from the erosion of these beach sediments and then their burial on that erosional face by the lower-water beach development during the neap tide. The deformed beds at the top of the section just under and to the right of the scale are footprint-deformed laminae. Photo courtesy of Miles Hayes.

The typical beach surface antidune pattern is seen to the left of the scale. The alternating light and dark stripes are a result of the separation of the heavier black mineral grains and the lighter-colored light minerals. In the very top of the trench, the landward inclination of the thin laminae of the truncated antidunes is barely visible. Deeper in the trench the alternate light and dark beds reflect times of normal beach conditions (light beds) and stormy periods (when the dark-colored heavy minerals were deposited). Photo courtesy of Miles Hayes.

waves. Such cross beds help define deposits that formed just offshore, on the surface of sandbars or tidal flats.

As noted in chapter 6, sheltered tidal flats and the low troughs between the swash zone and the inner sandbar are usually covered with fields of ripple marks that vary in their degree of asymmetry and occur in varied orientations, often with more than one set of ripples interfering with one another. When buried, these variations give rise to complex patterns of bedding that are a challenge to unravel when reconstructing the events that created them. Similarly, the antidunes that give rise to the stripes on beaches are expressed in the subsurface as cross laminae.

BURROWS AND BIOTURBATION

The animals discussed in chapters 9 and 10 that leave their tracks and trails on the beach also produce unique structures below the surface. Their activity comes after the physical processes that deposit the sediments, so their structures are said to *crosscut* the existing laminae and bedding. Sometimes there is so much organic activity just below the surface by burrowers and sediment reworkers searching for food, ingesting food,

Different types of burrowing and bioturbation are revealed in this trench, including burrows that are vertical, crosscutting up through the laminae; burrows that are probably near horizontal, such as the circular cross section at center bottom; and a zone or bed of completely disturbed, or bioturbated, sediment in the middle of the section. This latter bed was homogenized or uniformly mixed by the burrowers, probably as they searched for food. In contrast, the vertical burrows either are where an animal was digging its shelter or are escape burrows of an animal buried by beach-sand deposition. Photo courtesy of Miles Hayes.

and excreting sediment that the physical features of sedimentary structures are highly altered or destroyed. The sediment is said to be *bioturbated*, a sort of natural homogenization, or mixing. Even plant roots, such as those found in dunes, may have this effect. Bioturbation typically occurs in those parts of the beach where lower wave energy favors more organic activity. The sediments beneath the calmer waters over muddy tidal flats, nearshore troughs sheltered by sandbars, back-island lagoons, and salt marshes are more likely to be bioturbated than most parts of the open-ocean sandy beach. Nevertheless, those burrowers and subsurface feeders adapted to the beach environment will leave their traces in the subsurface.

Individual organisms produce unique burrows or subsurface traces that allow their identification even in their absence. Crabs are a good example on the upper beach. Burrowing ghost shrimp may characterize the low-tide beach. Species of the genus *Callianassa*, common in much of the world, typically live in tidal-flat sediments as well as those just landward of the offshore bar, and their burrow openings (see chapter 9) are easily identified. Some careful digging with a shovel reveals that these burrows, made from mud and organic particles scavenged by the shrimp, can descend well below the surface and branch in different directions. In contrast, the higher-energy swash zone is less likely to show burrower activity. Even here, though, the activity of a larger organism may deform the subsurface structures. Human footprints, tracks of turtle crawls, and the wallowing tracks of seals and sea lions distort the underlying laminae. Where vehicles are permitted on beaches, their weight will compress the layers beneath the tire tracks.

EVIDENCE OF A RETREATING BEACH

The cross-sectional view provided by a shallow trench in a beach is too limited to reconstruct the long-term history of the shore. For this broader view, a cross section of much greater depth and length is needed because the entire beach system,

Tree roots sticking up through salt marsh peat, on Jasper Beach, Maine. Below the tree roots is an upland soil. The sequence in which the layers formed involved a terrestrial setting in which a spruce tree germinated. At some point the ocean encroached on the tree and it died. A salt marsh eventually covered the tree roots, and the marsh persisted for possibly hundreds of years before gravel from the beach was thrown on top of the marsh, killing it. More gravel piled up (20 ft [6 m] thick), and over thousands of years the entire beach migrated over the spot. Now the site has been eroded on the seaward side of the spit. The tree fragments look relatively fresh, but the peat has been flattened greatly by the passing weight of the beach. The knife is shown for scale.

Geologists from the University of Maine coring through the lagoon deposits on the landward side of Jasper Beach spit. The pipe is vibrated into the lagoon mud and then retrieved, providing a continuous core sample from which the history of sedimentation can be determined.

barrier island, or coastal zone is layered as well. We get an occasional glimpse of this stratigraphy, or layered sequence, at sites where storms or long-term erosion have exposed buried peat layers or tree stumps in the swash zone. Clearly a tree cannot live on an ocean beach, and mud layers rich in organic plant roots do not accumulate on wave-washed beaches. These exposures must represent a time and place when the shoreline was in a more seaward position and sheltered forest or back-island salt marsh grew in the position now occupied by the beach.

Geologists have long recognized that the swash-zone appearance of fossil plants and animals that once lived in a lagoon behind a beach meant that the beach had retreated landward from a more seaward position and had rolled over the lagoon deposits (see chapter 10). Rising sea level and diminished sand supplies are accepted as the ultimate driving forces causing landward beach migration, although storms are usually the immediate reason we find peat layers uncovered on the beach. Wanting to be sure of this and to determine the rate of landward beach re-

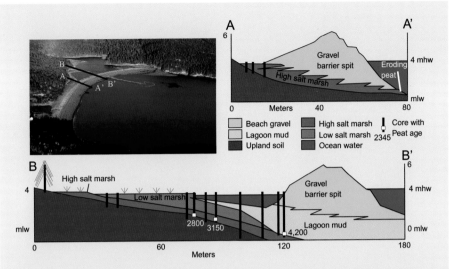

The cross section of Jasper Beach, Maine, developed from core samples and exposures on the beach. The photo of the core being taken was along line B–B' and contained a peat sample that gave a radiocarbon date of 3,150 years before present (mhw, mean high water; mlw, mean low water).

treat, coastal scientists core through beaches and dunes, usually by vibrating long pipes into the beach to extract long core samples of its interior. These cores are taken along a line across a barrier island and are used in conjunction with ground-penetrating radar or seismic surveys, or both, taken along the same line, to identify the various buried layers, correlate them laterally, and obtain samples of the sediments. If the cores penetrate buried salt-marsh muds and peats that include organic materials such as plant roots and shells of organisms, the organic material in the core can be sampled and dated using the radiocarbon method. The ages of the peats indicate how long ago the salt marsh was growing in its position behind a barrier island, and how long the beach and associated dunes have been moving landward to bury the old marsh. In this way, the rate of the sea-level rise can also be determined. Similarly, the cores are used to verify the various sedimentary units that show up in the seismic or ground-penetrating radar records. For example, former inlets that once cut through barrier islands and then were completely filled with sediment can be identified by the unique radar or seismic patterns of those beds that accumulated on the flanks of the inlets and are inclined, or dip, toward the former inlet channels.

PART III

THE GLOBAL THREAT TO BEACHES

Beaches are dynamic systems that have been important to humans throughout history. They remain one of the most economically important natural environments, providing both essential habitat in the supporting food chain of fisheries and keystones to recreational economies. These sand and gravel ribbons are the land's protective cushions against the impact of storms. We should all treasure beaches, and the preceding chapters have sought to improve the reader's understanding of their beauty, mystery, and fragility. But beaches are and have been under human threat for the last century or so.

Chapters 12 and 13 turn from the sunny, fun-at-the-beach scenario to the reality of our need for beach conservation. Even frequent beach visitors are surprised to learn that many of the world's best-known beaches are now artificial, no longer holding their original composition or their natural configuration. Waikiki, Hawaii; Rio de Janeiro, Brazil; Miami Beach, Florida; the French Riviera; Durban, South Africa; Surfers Paradise, Queensland, Australia; Barcelona and Marbella, Spain; and Venice Lido, Italy, are only a few of the famous beaches that have been replaced by nourished replicas. At Waikiki the nourishment sand has been imported from a neighboring island and from as far away as the U.S. West Coast. At Lyme Regis, England, 75,000 tons of sand were brought from France. Nourishment sand is a limited resource, and the cost for such soft solutions (as well as that for hard solutions) is rising. Likewise, the uncertainties of both natural and political environments raise serious global questions.

Are beaches a threatened species? Certainly beaches in the geologic past moved back and forth as the level of the sea changed. The beaches survived just fine, but now things are different. We're here. We've jammed buildings up against beaches just about everywhere, it seems. Now we have to ask such questions as: What are the effects of sea-level rise and climate change on beaches? Are buildings more important than beaches? Do we want engineered systems that provide for temporary economies and keep buildings in place, or do we want natural sustainable beaches at the eventual price of losing the buildings?

12

BEACHES AT RISK: SEA-LEVEL RISE
AND THE HUMAN RESPONSE

Nature, to be commanded, must be obeyed.

FRANCIS BACON, *NOVUM ORGANUM*

The single greatest threat to the future of the world's beaches is not storms or rising sea level. It is us. The threat comes not from nature, but from humans in our attempts to control the beach. Beaches, left to their own resources, are extremely resilient. As a rule, natural beaches are nearly indestructible; indeed, it appears that beaches have discovered the fountain of youth. Whatever the level of the sea, beaches will persist. Every rule, of course, has exceptions, and in a few instances beaches can meet their demise by natural circumstances (e.g., small beaches in non–coastal plain settings may run up against rock cliffs, the natural sand supply may be cut off, or the relative sea-level rise may be rapid enough to overwhelm the shore, as in the case of rapid subsidence due to an earthquake or a sinking, abandoned delta). In most cases, however, even in big storms, the beach survives and, given time, repairs itself from storm damage. The media may say a particular hurricane or storm has caused "beach erosion" or "severe beach loss," but a visit to the shore tells us the beach is still there (perhaps sans buildings).

All over the world, we have crowded buildings up against the shoreline, right next to beaches that are widely recognized to be retreating (migrating landward). But instead of moving landward with the rising sea level as they naturally would, these beaches are forced to stay in place in order to protect the buildings. Clearly some people value buildings over beaches and are often willing to protect buildings even at the price of the loss of the beach.

A wide beach masks the spectacular events that have taken place on appropriately named Washaway Beach in the U.S. state of Washington. A careful look at the photo reveals four pipes protruding from the beach, lined up in the offshore direction. Each of these pipes extended to a well from a house that has long since fallen victim to storms and beach retreat. Over the last decade or two, four houses have fallen here.

The biggest long-term threat to buildings next to the shoreline is sea-level rise. The sea level is rising because the surface waters of the oceans are warming and expanding, and ice sheets and glaciers are melting. In the early twenty-first century, the rate of global sea-level rise was a bit more than 1 ft (0.3 m) per century. All indications are that the ice sheets' melting rate is increasing and that sea-level rise is accelerating. We believe that 3 to 5 ft (0.9 to more than 1.5 m) of sea-level rise should be expected by the year 2100. In fact, a 7-ft (2 m) rise is not out of the question.

Three feet (0.9 m) of vertical sea-level rise does not sound spectacular, but for low-lying, gently sloping coasts characteristic of the world's coastal plains in China, Brazil, and the United States, the resulting displacement of the shoreline will be far inland. Three feet of sea-level rise will be enough to halt all development on the world's barrier islands as we know them. Three feet will flood much of the city of Miami, Florida, and large portions of other cities such as Rio de Janeiro, Brazil; Venice, Italy; New Orleans, Louisiana; Durban, South Africa; Lagos, Nigeria; Gold Coast, Australia; Tel Aviv, Israel; and Cadiz, Spain,,to name only a few. Low-lying island nations already are feeling the consequences of the sea-level rise (e.g., the Maldives in the Indian Ocean; Tuvalu, Kiribati, and others in the Pacific Ocean). Outside of the cities, a worst-case scenario is that many of the world's beaches that are accessible to tourists will become engineering projects. Most beaches will become roads for bulldozers spreading newly placed sand or constructing seawalls in an attempt to hold back the sea to protect buildings, foolishly placed next to the ocean.

The only "solution" to the erosion problem that minimizes harm to the beach is to move buildings back from the shore, demolish them, or let them fall into the sea. This, the so-called retreat option, is routinely practiced in the developing world but is less attractive in the world of wealthy and politically influential beachfront property owners, who opt for engineering solutions instead.

COASTAL ENGINEERING

The "we shall not be moved" approach confronts the sea-level rise with either or both of two common engineering responses: *hard stabilization* or *soft stabilization*. Both are very damaging to beaches, especially the shore-hardening approach.

HARD STABILIZATION

Since the time of the ancient Romans, humans have attempted to engineer shorelines by building hard structures to hold the shoreline in place. Hard stabilization consists of putting something hard and "permanent" on the beach. Most common are *seawalls* (and their cousins, *revetments* and *bulkheads*), which are walls built parallel to and on the landward, or back, side of the beach. A seawall may be a wall of concrete or steel; a line of rocks, logs, or sandbags at the top of the beach; or a massively engineered structure such as the walls in Galveston, Texas, and along the coast of Belgium, the coastal defenses surrounding much of Holland, and the walls on the barrier islands protecting the Venice lagoon and the city of Venice, Italy. Many esplanades on urban shores are in effect glorified seawalls, including well-known examples such as Brighton, England; Nice, France; Rio de Janeiro, Brazil; Tel Aviv, Israel; Beirut, Lebanon; and Colombo, Sri Lanka.

In the United Kingdom, seawalls are particularly common in resorts, where many were built to support concrete promenades rather than as

This motel at South Nags Head, North Carolina, extends beyond the midtide line on the open-ocean beach. One of the first medium-rise buildings on the Outer Banks, it was built well back from the beach, but a steady erosion rate of 3 ft (1 m) per year has moved the beach into this position. More than forty rooms now have a beautiful view of the sea, although one might wish to check the weather forecast before renting one of them.

These buildings along the much narrowed beach in front of the Mar Menor, Spain, seawall are in a hazardous zone. The round columns on the beach are part of the sewer system that has been exposed by shoreline retreat.

sea defenses. The seawall in Blackpool often is topped by waves during storms. The Dawlish seawall protects the southwest coast railway line, and trains frequently have to stop running during storms. The possibility of rerouting the southwest railway line has been discussed, but it was ruled out by local government. The seawall from Hopton in Norfolk to Corton in Sussex was closed to the public in 2008 after a storm damaged the wall.

The remnants of sandbag seawalls litter the beach after the passage of Hurricane Ivan on Galveston Island, Texas. Sandbag seawalls are often viewed as a compromise solution to be used instead of concrete. The problem here is that such remnants are seldom cleaned up, leaving behind a highly degraded beach after the sandbag seawalls are destroyed.

Three generations of sand bags are visible on the beach at Figure 8 Island, North Carolina. The oldest set is the white fragments visible in the foreground. The second sandbag seawall was made up of the black sandbags that can be seen in the middle of the photo. At the time the photo was taken, a third sandbag seawall was located up against the dune scarp.

Destruction of the beach in front of a seawall is largely a passive process that can take years or decades. The wall is constructed because the shoreline is retreating, but the wall does not address the wave and current processes that are causing the shoreline to retreat. Thus, after the wall is put in, the beach continues to retreat, getting narrower and narrower as it backs up against the wall. Eventually the dry beach disappears.

This process usually takes a few decades, but it can happen much more quickly. In the United States, this fact has been recognized by some, and Texas, Oregon, South Carolina, North Carolina, Rhode Island, and Maine have all taken steps to limit and even outlaw seawalls and other hard structures on their beaches. But politics in democratic societies often favors development over environment, and it is difficult to halt the march of seawalls along developed beaches.

In the United States, the state of Florida is in the worst shape of all and provides a good illustration of the secondary effects of seawalls. The walls provide a sense of security, which encourages construction of high-rises. It is one thing to move cottages back from a rising sea level and retreating shoreline, but a twenty-story condominium is another matter. The point is, large buildings will precipitate heroic attempts (that will inevitably fail) to hold the shoreline in place, and meanwhile the beach will be history. The state of Maine recognized this effect twenty years ago and banned all buildings taller than 35 ft (10 m) from coastal sand dune areas.

A very crowded, narrow pocket beach in Estérel, France. Beach volleyball was invented in France, but the court would be narrow for a game here. Presumably the seawall is responsible for the narrowed beach.

Another approach to hard stabilization is construction of walls built perpendicular to the shoreline called *groins* (*groynes*). These structures are designed to trap sand that is transported along the beach by longshore currents and thus widen the beach. Groins range in length from a few tens of feet to a few hundred feet and are successful in widening a beach when there is sand available. The problem is that the shore to which the trapped sand was going (the adjacent beaches) is now suffering a sand deficit, which of course leads to increased erosion rates there. Perhaps the most spectacular damage done to beaches by groins occurs along much of Portugal's open-ocean coast north of Lisbon. Most of these groins were put in place less than three decades ago.

Groins are just about everywhere on the south coast of the United Kingdom. Brighton, Bournemouth—all the main seaside resorts have them. In fact, groins have become a part of the traditional English seaside resort landscape. Interestingly, in Barcelona, Spain, the local government created an unexpected uproar when it removed some groins and nourished the beach instead. It turns out that the groins had divided the beach into de facto zones used by families, the gay community, the nudist community, and so forth, and their removal had removed the zones!

Jetties are a generally larger form of groins that are placed adjacent to tidal inlets and river mouths to keep sand out of navigation channels. These structures also have the potential to trap huge quantities of sand and wreak havoc on the downdrift beaches that would have received the trapped sand. Just that has happened in several well-known U.S. examples, including Ocean City, Maryland and Charleston, South Carolina. It is estimated that half of the erosion of Florida's beaches results from jetties at inlets. In some locations, where the inlet is also the mouth of a river that delivers sand to the beach, a jetty can cause erosion by depriving the adjacent beach of river sand. Camp Ellis, Maine, is an example of that sad situation. Such long-term erosion effects also are evident in many United Kingdom and European examples, where jetties have

Upper left This seawall on the Majuro Atoll, Marshall Islands, is made of gabions, wire baskets filled with rocks and shells. The seawall holds back the landfill of the town of Majuro. These gabion wire baskets are notorious for corroding within five to seven years, rupturing and spilling their contents on the beach and becoming a hazard to walking on the beach. If not replaced, their failure will allow the refuse to escape, creating a pollution hazard. Thousands of coastal landfills around the world are threatened by the sea-level rise.

(Captions continue on facing page)

Upper right This unique version of a seawall in French Polynesia is made of palm logs, facing the open ocean on an atoll. Obviously this wall was made from local materials! Note the concentration of coral and associated carbonate sediment derived from the offshore reef (white strip). The high tide regularly reaches the wall, which offers little protection against a big storm.

Center upper left An eroding landfill on the Puget Sound shore, Port Angeles, Washington. The shoreline was walled in 2006 to halt the erosion (for a while). Courtesy of Hugh Shipman.

Lower right The Galveston, Texas, seawall probably is the mightiest seawall on any barrier island in the world. Constructed after a 1900 hurricane that killed six thousand citizens of the city, the wall caused the eventual narrowing and disappearance of the dry beach as the beach and offshore slope steepened. A rock revetment was placed at the toe of the wall to protect its foundation. As steepening continued, a second rock revetment was put in to protect the first rock revetment.

Center lower left This seawall at Salinopolis, Brazil, just south of the Amazon River mouth, was flanked by accelerated erosion at the end of the seawall, rendering it useless. Instead of protecting property, the wall has caused littering of the beach and contributed to the shoreline erosion hazard. Seawalls, once installed, always require extensive and expensive maintenance.

Lower left This seawall in British Columbia, Canada, was built from logging-operations debris. As is commonly the case, easily available local materials are used to make seawalls, but no matter the type of materials, on eroding shorelines, seawalls cause beach loss.

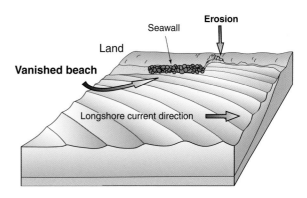

A diagram of the effects of a seawall. A rigid wall reflects and refracts wave energy, causing erosion in front of the wall and concentrating energy at the ends of the wall, also resulting in erosion. The net effects are beach narrowing, beach steepening, and the ultimate loss of the beach in front of the wall; breaching at the end of the wall; and the cutoff of the sediment supply to downdrift beaches. Drawing by Charles Pilkey.

been in place for very long periods of time. Offshore *breakwaters* are a third species of hard-engineering structures, consisting of a wall built parallel to the shoreline, but these are positioned offshore a few tens or hundreds of feet. These structures reduce wave energy reaching the shoreline by creating a wave shadow on the beach, interrupting the longshore current, which causes the wave-transported sand to stop and accumulate. In other words, a breakwater has an impact similar to that of a groin, in that it causes the beach to widen at one point at the expense of the downdrift beach, which narrows or disappears.

Unfortunately, large storms overwhelm breakwaters and damage or destroy the buildings the breakwaters were designed to protect. Holly Beach, Louisiana, hosted a vast replenished beach and offshore breakwaters far more costly than the beach houses they were built to save on the "Cajun Riviera." When Hurricane Rita struck nearby in 2005, the breakwaters withstood the storm quite well, but the sand and houses were virtually all lost.

A field of natural groins is the result of low rock ridges caused by outcropping rock layers that strike perpendicular to the beach. Sand has accumulated on one side of the natural groin (middle of photo) and caused the shoreline to retreat on the other side (background). The only individuals impacted by this sediment deprivation would be the beach mammals and birds that frequent this beach on the Antarctic Peninsula. Photo courtesy of Norma Longo.

A good example of offshore breakwaters in the United Kingdom is at Sea Palling in Norfolk. In the Blackwater Estuary in southeast England, a line of sixteen old barges was sunk about 550 yd (500 m) offshore to create some very unusual offshore breakwaters.

Some confusion exists when discussing shore-hardening structures because of differences in terminology between countries. Jetties, as defined above, are called training walls in Australia, while they use the term jetty for what are termed recreational piers in North America and the United Kingdom. In New Jersey, groins are sometimes called jetties. But regardless of the name, these are all hard structures that impact shorelines.

All in all, if preservation of beaches for future generations is a high priority, all forms of hard stabilization should be avoided. This means of course that, in a time of rising sea level, buildings must be removed and the shoreline must be allowed to march onward. This option is unreasonable in an urban setting. Cities such as Tokyo,

Singapore, and New York City are not going to retreat as the sea level rises. The priority in an urban situation is preservation of the buildings and infrastructure, but along most of the nonurban ocean shoreline, we believe the priority should be preservation of beaches and not the defense of beachfront houses.

This groin in Portugal clearly has blocked the sand movement from left to right. In Portugal, installation of a number of such groins along the Atlantic coast created erosion problems along more than a hundred miles of shoreline. Photo courtesy of Miles Hayes.

SOFT STABILIZATION

Although the designation *soft* may be misleading, soft stabilization refers to engineering designs to hold beaches and dunes in place by artificially adding sediment or, in the case of dunes, using vegetation to trap and hold sand. Most commonly for beaches, this approach involves the construction and maintenance of artificial beaches by the process of *dredge and fill* or by hauling in sand to place on the beach, referred to as *beach nourishment* or *beach replenishment*. Although marketed under the pretext of protecting beaches, this soft approach again is usually aimed at protecting property (buildings and infrastructure), not the beach habitat. Nevertheless, this approach has become the solution of choice for many retreating beaches in the developed world. Construction of artificial beaches, however, is simply too costly for the developing nations.

Such replenishment is carried out either by dredging sand from the seafloor and pumping it to a beach or by transporting sand with trucks from inland sources. Dredging is usually carried out in lagoonal waters, in tidal inlets, or on the adjacent continental shelf. The cost in the United States ranges from $1 million to $10 million per mile of open ocean beach. On a volume basis, the U.S. cost of sand ranges from $2 to $40 per

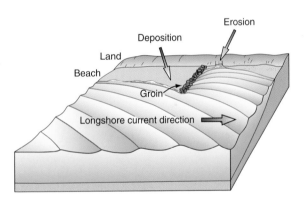

Diagram of the effects of groins. Groins block the natural alongshore drift of sand, cutting off the sand supply to downdrift beaches and contributing to their erosion. Several groins are usually placed along a beach (groin field), magnifying the effect. Often they are ineffective during storms, breaching on the landward end and losing their trapped sand. Drawing by Charles Pilkey.

A view of groins from a cliff top in Lima, Peru, shows sand moving from the top to the bottom of the photo. The narrowed beach in the foreground is starved of sand and shows well-developed beach cusps (bottom center) and an erosional scarp at the back of the beach cutting into the upland, which eventually will threaten the road.

cubic yard, and the volume of sand per mile of beach is usually between 150,000 and 1 million cu. yd. (between about 115,000 and 765,000 m³). The life span of nourished beaches is quite variable, ranging from days (Cape Hatteras, North Carolina) to as many as two decades or so (Miami Beach, Florida). Most replenished beaches on the U.S. East Coast have life spans of between three and five years.

The Dutch require that sand for beach nourishment of the Frisian Islands must be obtained from water depths of 65 ft (20 m) or more. This means that the North Sea sand is dredged from over the horizon. This distance is important because if depressions are dredged on the continental shelf too close to the shoreline, they will alter wave patterns on the beach, often leading to accelerated erosion in unwanted locations. Elsewhere in the world, including sometimes in the United States, dredging is frequently carried out within a matter of a few hundred yards of the beach, often contributing to rapid loss of the artificial beach (as happened in Grand Isle, Louisiana).

Another problem is with the choice of sites from whence to obtain sand. On barrier islands, the highest-quality and cheapest sand is found in inlets. The sand in inlets is beach sand that reached the inlet by longshore currents. The problem with mining inlets is that the hole that is left behind rapidly fills with sand that would otherwise have crossed the inlet to the beaches on adjacent islands. Thus, mining sand in inlets to halt erosion ultimately causes increased erosion on nearby beaches.

Surprisingly, after a half century of beach-nourishment projects around the world, there are relatively few studies of the impact of such projects on the biota. Beaches and their associated inner shelf areas are part of the nearshore ecosystem. The construction of artificial beaches certainly impacts beach fauna during the construction phase, and there are a limited number of studies to suggest that the impacts are longer term, perhaps permanent in some cases.

The Five Sisters (five breakwaters) protect Winthrop Beach, Massachusetts. The natural beach disappeared long ago, and large seawalls and groins protect property along with the breakwaters. The breakwaters create a "shadow zone" of lower wave energy, and tombolos are forming. These sand bodies are accumulating from the shore to the breakwaters and almost connect the breakwaters with a replenished beach. The tombolos in turn prevent any longshore transfer of beach sand from behind this structure, cutting off the sand supply to the downdrift beach.

The food chain of the beach begins with the debris washed up on the shore, and the nutrients carried in water from the ocean and the land. These nutrients sustain the microscopic animals that live between the sand grains and go up through a complicated web of feeders including crabs and mollusks to the birds and fishes that feed on them. Each level in the ecosystem depends on all the others. Adding to the mix are the small mammals that scavenge whatever washes up on the beach and, where possible, eat the eggs laid by the turtles and birds on the beach. The lions that survive on the Namibian desert coast by eating dead whale carcasses are one of the most bizarre examples of beach scavengers.

Needless to say, the animals that live on and in the beach are used to extreme events such as storms. The living things in the beach are frequently moved about, buried, and unburied, and somehow they manage to survive. However, beach nourishment creates a different set of stresses on the organisms. On a large nourished beach, most living things are killed when they are buried by a yard or more of sand in a very short period of time. Eventually the organisms come back, reestablished from populations from adjacent beaches, as long as the new sand on the artificial beach is about the same size and composition as the original beach. If the new grain sizes and composition are not compatible, a whole new type of fauna may arrive. If the nourished sand has a lot of mud, the new fauna will be very different indeed. Recovery of the living things in an artificial beach depends on their mobility, but complete recovery probably takes several years.

A grounded supertanker off Goa, India, is acting as an offshore breakwater, causing the local beach in front of the restaurant to build out. At the time this photo was taken, the tanker had been grounded for more than a year. Photo courtesy of John Gunn.

A cascade of sand slurry (the pipe mouth is in the sand mound) being pumped onto a beach at Durban, South Africa, for beach nourishment. The sand is dredged from the adjacent port entrance, where it accumulates as a result of longshore drift from the south, and then pumped to the beach. This operation of bypassing the sand across the channel is carried out once a year, every year. The young man in the foreground is probably looking for seashells.

Also, many of the areas impacted are offshore, out of sight, where invertebrate populations live on so-called hard grounds, where meadows of sea grasses thrive, where fish nest and feed, and, in tropical regions, where coral reefs need clear waters to thrive.

The biological impacts of nourished beaches also involve the shorebirds that nest on the beach. These include various plovers, sandpipers, terns, and skimmers. Studies have shown that an artificial beach changes foraging behavior, leading to more feeding time and less food. Decreased parental time leads to predator problems for the newly hatched chicks. The density of nests usually declines dramatically, as nesting shifts to other beaches. Turtles too have problems on nourished beaches. They sometimes can't dig through the compacted sand and can have their flippers cut to shreds by sharp shell fragments broken up in the dredge hopper.

The quality of sand used to make an artificial beach is one of the biggest issues in coastal engineering. All over the world there are examples of nourishment sand that is of poor quality—too muddy, too rocky, too shelly. A number of Spanish replenished beaches are reported to be muddy sand. In the state of North Carolina, which has laws requiring "compatible" sand for artificial beaches, there have been artificial beaches that are rocky (Oak Island), shelly (Emerald Isle), full of construction debris (Holden Beach), and muddy (Atlantic Beach). In all cases, the site for obtaining the sand was poorly sampled in an effort to save money, and government failed to halt the emplacement of poor-quality sand. In the rush to save the beachfront property, it seems anything goes.

Beach bulldozing or *scraping* is another soft method of protecting buildings. Sand is bulldozed or scraped from the low-tide beach up to the back of the beach to form a "storm dune" (an artificial sand dike) to provide temporary protection to buildings during minor storms. This approach provides no long-term protection, however, because no new sand is added to the beach. The scraping has the added disadvantage of destroying the beach fauna.

This close-up view of a shell hash on a newly nourished beach on Emerald Isle, North Carolina, shows how nourishment can sometimes degrade a beach. The original beach was composed of a more uniform sand with a pleasant tan color (the area is known as the "Crystal Coast"), but the new beach was very coarse due to an abundance of broken shell material, and dark gray in color because of the high content of black shell material. The initial new beach was made up mostly of shell hash, including much larger shells than shown here, and the broken shells had sharp edges. The beach was no longer a pleasant barefoot experience. With time, the finer sand winnowed out of this material to cover the shell hash with a thin layer of sand, but in the intertidal zone that people cross to go swimming, shoes were required because of the abundant sharp shells. The dark color also lessened with exposure. Courtesy of Adam Griffith.

OTHER DAMAGING ACTIVITIES

Beach grooming is a common procedure on recreational beaches to make them more pleasant for tourists, but this is another damaging process in

This erosion scarp on a nourished beach at Kashima, Japan, exposes shell hash and dark-colored sediment. Scarps are almost always present on nourished beaches, a sign that the new beach is eroding much faster than the previous natural beach. A clue that this beach is artificial is provided by the fact that all of the shells are white and mostly fragmented.

terms of the impact on the tiny creatures that live within the beach. In addition, beach grooming may contribute to shoreline erosion. One hotel owner on the island of Roatán, Honduras, indicated that the elevation of the hotel's beach was lowered by 1 ft (30 cm) after a few years of daily raking. About half of all of Southern California tourist beaches are groomed once a day, and the urban beaches of Los Angeles are often raked twice a day to pick up garbage, cigarette butts, cans, and whatever else people leave behind. Usually such grooming is done by raking, but sometimes it is carried out by sieving the sand. In either case, the living organisms are largely removed. Even dune-building plants that may be creeping out onto the beach from nearby dunes are routinely killed. A bold Least Tern that chooses to nest in the most seaward clump of dune grass will be a potential victim of overzealous beach grooming. In any case, the beach becomes a sterile desert. One solution is to alternate groomed and nongroomed areas or to hand clean the beaches, perhaps using volunteers who love beaches and who love to stroll on them. Some Belgian beaches are now routinely hand-picked, specifically to remove garbage but leave natural beach debris behind.

Invasive and alien species can result from activities that are well-intentioned or deemed harmless. Artificial plantings of nonnative plant species to stabilize dunes are an example noted earlier. Almost every dune system in the world has examples of plant species that originally were introduced into flower gardens but then spread into the dunes at the expense of native species. In coastal regions in the Caribbean and southeastern United States, Australian pines have been planted as windbreaks and for their aesthetics, but these trees grow to the edge of the back of the beach and cover important nesting grounds for birds in particular that would be attracted to the otherwise open sand flats. Australian pines have shallow root systems and are notorious for blowing over in hurricanes and blocking roads as well as littering back-beach nesting grounds.

Several alien animal species have negatively impacted North America's Great Lakes, but none more so than the zebra mussel. These prolific small clams have clogged water intakes and outfalls, altered water quality, and grown to such abundances that in some places their shells have become the dominant material on local beaches.

Beach driving, or vehicular traffic on beaches, deserves special mention. Whether or not to allow vehicles on beaches is a contentious issue in many countries. Daytona Beach, Florida, is famous for its liberal beach-driving policies, which have been in place for more than one hundred years. The Texas Open Beaches Act acknowledges a right of people to drive on the beach that is greater than the right of people to have a house on the beach. In northwest Europe, where there are many wide beaches with hard surfaces of fine sand, there is a long tradition of driving normal two-wheel-drive cars onto the beach. Beach parking lots are even provided, and a fee is sometimes levied for car access to the beach. In South Africa, as mentioned earlier in the book, beach driving was recently halted in the entire country, and within two years beach life had sprung back very impressively. Australian beach managers are working to restrict beach driving in that country to protect and restore habitat.

Post-storm beach bulldozing (beach scraping) on a hazy morning at Kitty Hawk, North Carolina. The sand is pushed up to the beach cottages to provide a buffer against the next storm, but such sand dikes (they are not dunes) are quickly whisked away in storms. The seagulls are gathering for the free lunch provided by the bulldozer as it exposes (and destroys) the beach ecosystem.

The problem with beach driving is that heavy traffic does essentially the same thing to the beach fauna that grooming does. People who fish and surf are the main culprits, but sightseers and those who just want to try out their four-wheel-drive vehicle are among those crowding beaches. In addition, some "outlaw" drivers wander into the dunes, killing nesting birds, destroying beach vegetation, and destabilizing dunes. Beach driving is a practice that should generally be opposed if the quality of beaches and dunes is to be maintained for the majority of beach users.

WATER POLLUTION

The problems of beach pollution are widely recognized around the world. The most obvious major pollution source comes from untreated or partially treated sewage. Often treatment plants are overwhelmed in heavy rainfalls, and raw sewage is added to the stormwater runoff. In beach communities, malfunctioning septic tanks are a big problem.

Workers grooming the beach in the Maldives, Indian Ocean. The use of rakes is certainly better for the beach fauna than tractor-drawn and heavy mechanized beach sweepers. Nonetheless, such beach cleaning disrupts the beach microfauna, creating breaks in the food chain that lead all the way up to the fish cruising in nearshore waters. It is better to hand pick the trash from the beach.

This pile of refuse on a Bali beach in Indonesia is the result of an effort to clean trash from the beach. Globally, such scenes are common as beaches accumulate wrack from human sources, usually dominated by plastic. Places like Bali and Sumba Island still have beautiful beaches, but they are threatened by solid waste, as seen here. Photo courtesy of Claude Graves.

Australian Pines line the back of this Bahama Islands beach, unusual in that they grow almost to the high-tide line, so they commonly fall to erosion and accumulate in the back-beach area. The trees are a problem for tourist recreational beaches and a serious problem for birds that nest on the upper beach. The Australian pine here and in much of the Caribbean and south Florida is an invasive species. Photo courtesy of Sidney Maddock.

Dangerous conditions in beach swimming areas can be determined by analysis of the water for certain pathogens and indicators (e.g., E. coli bacteria). Certain organisms, including some types of seaweed, are also indicators of pollution. Off of Adelaide, Australia's shore are areas of extensive sea grass loss, caused by industrial and wastewater discharge. This loss has been directly correlated to an increase in the volume of sand on the adjacent beach.

A partially buried car on Washaway Beach, Washington, leaves one wondering how it got there. A look to the background tells part of the story: a bare scarped bluff face, downed trees, and remnants of an infrastructure. This beach is eroding at more than 100 ft (30 m) per year, and we can conjecture that the car's fate was due to a storm and shoreline retreat that caught up to a house and driveway.

It is important to remember that on polluted beaches, building sand castles with wet sand and other beach activities can be just as hazardous to one's health as swimming in the polluted water.

One way to ensure that the beach where your family swims is safe is to use Blue Flag Beaches only. The *Blue Flag Beach Programme* of the Foundation for Environmental Education (www.blueflag.org) has approved more than thirty-three hundred beaches in twenty-nine countries in Europe, the Caribbean, Canada, New Zealand, and others. The requirements for approval include good management and good water quality.

In the United States, the Natural Resources Defense Council publishes an annual report (www.nrdc.org/water/oceans/ttw/ttw2009.pdf) on the quality of beach water at two hundred of the most popular beaches in the United States. The primary causes of pollution are storm-water runoff and sewage release. More than twenty thousand brief closures occurred at these beaches in 2008. The worst U.S. beaches were in the Great Lakes and the best were along the shores of the southeastern United States.

OIL SPILLS

On April 20, 2010, the mile-deep BP Deepwater Horizon oil spill began in the Gulf of Mexico when the blow-out preventer on the sea floor failed and the oil rig exploded. This spill exceeded the *Exxon Valdez* disaster to become the biggest oil spill in U.S. history. The well was finally sealed on September 19, 2010, after gushing close to five million barrels of oil into the Gulf. The impact of this spill will be felt for decades to come, and it is the harbinger of spills yet to come.

Upper Tire tracks on beach sand rich in heavy minerals. The tracks are rather striking because they were made on a beach surface with a thin layer of white sand underlain by a thicker layer of black heavy-mineral sand. Vehicular traffic on beaches damages bird and turtle nesting areas and also impacts the microfauna within the beach— damage that goes unseen.

Lower Tire tracks on the beach at Doñana, Spain, are deep and abundant. Designation of such an area as a park has little effect in terms of conservation if driving on the beach is allowed. No successful nesting will take place here, and the beach fauna will be sparse as well.

Many beaches in the world have small tar balls waiting in the sand for the unwary stroller to step on, blackening the soles of their feet. Some resorts along Mediterranean and Caribbean shores provide beachgoers with brushes and cans of solvents to remove oil from bare feet upon leaving the beach. Tar balls, some as small as sand grains, are the end product of oil spills after all the light components of the oil have evaporated away. Most are the result of small amounts of oil released into the ocean by passing ships, a practice that has decreased dramatically in recent years.

Oil spills, especially the largest ones, which occur as a result of accidents with tankers and wells, provide the most spectacular pollution incidents for beaches. There are two kinds of oil spills: chronic and episodic. Chronic spills are those that occur frequently in the vicinity of ports where there is ship traffic. Episodic spills are the occasional and relatively uncommon (at any given location) release of oil after an accident of some kind, often involving a tanker. The largest oil spill in history was the 1991 Gulf War oil spill in the Persian Gulf, which resulted from the purposeful release of at least 140,000 tons of oil by Iraq to stymie an expected amphibious landing by U.S. Marines. The ten largest oil spills in today's oceans as of 2010 are shown in Table 12.1.

The second largest spill in U.S. waters was the *Exxon Valdez*, which dumped 11 million gallons of oil into Prince William Sound in Alaska in 1989 and oiled portions of an 1,100 mi (1,770 km) stretch of shoreline, including sandy and rocky beaches. Although this spill ranks thirty-fifth on the all-time list by volume of oil released, it was one of the most damaging environmentally because of where it occurred. The oil residues from that spill can still be found at depth in the beach sediments, twenty years after the event. Similarly, the Kuwaiti and Saudi Arabian oiled beaches and tidal flats in the Persian Gulf have been slow to recover, and concentrations of oil are present at a foot or two below the surface. In contrast, the BP Deepwater Horizon spill initially appears to have been less damaging to

Table 12.1   The Ten Largest Oil Spills in Today's Oceans as of 2010

Event/Tanker	Location	Year	Gallons
Gulf War	Kuwait (Persian Gulf)	1991	Up to 520 million
BP Deepwater Horizon	Gulf of Mexico	2010	206 million
Ixtoc I Well	Gulf of Mexico	1979	139 million
Atlantic Empress	Trinidad and Tobago	1979	88 million
Fergana Valley	Uzbekistan	1992	88 million
Nowruz Field	Persian Gulf	1983	80 million
ABT Summer	Off Angola	1991	80 million
Castillo de Bellver	South Africa	1983	78 million
Amoco Cadiz	Brittany, France	1978	69 million
Amoco Haven	Genoa, Italy	1991	44 million

A TOXIC GREEN WAVE ON THE BEACH   by Sharlene Pilkey

With beaches all over the world already under attack from sea-level rise, mining, driving, seawall construction, and encroachment by human development, another menace is mounting an assault—and humans are behind this one too. Articles in the *Guardian* (August 10 and August 21, 2009) reported that lethal green algae have invaded heavily used vacation beaches in Brittany, northern France, and along England's coastline from Wales to Portsmouth. These deep piles, up to 3 ft (1 m) thick, with hard, crusted tops, are stinking, ticking gas bombs.

Vincent Petit, a twenty-seven-year-old veterinarian, was horseback riding on a Brittany beach near St. Michel en Grève, when his horse broke through the crust and went down. A cloud of hydrogen sulfide gas was released from the rotting algae, reportedly killing the horse within thirty seconds. Fortunately a tractor was nearby; it was used to clear away algae and drag Mr. Petit to safety. He was rescued in an unconscious state and hospitalized. He reportedly was suing the local municipality, which was responsible for beach maintenance.

On June 22, 2009, on the Côtes d'Armor, a forty-eight-year-old maintenance worker clearing dead algae from the beach was stricken and died, apparently from a heart attack, but the lethal rotting vegetation is suspected in his death.

This abundant accumulation of marine algae on the French coast was apparently the product of overfertilization of nearby fields and drainage emptying into the sea. Towns along the Brittany coastline have hired bulldozers to scrape the seaweed away, but more is brought ashore by waves.

Previously, on a beach near Mr. Petit's accident, two dogs met their death from the gaseous fumes. In a strange coincidence indicating the global nature of this problem, the death of two dogs running on an algae-encrusted beach was recently reported from the north of Auckland, New Zealand, as was the death of four dogs killed in 2009 by toxic beach algae near Elkton, Oregon.

The more one learns about this beach hazard, the more apparent its global scope becomes. In 2008, the Chinese government brought in the army to clear away the slimy green growths so the Olympic sailing competition could be held and so observers could safely view the event. In Italy, near Genoa, a sixty-year-old man had to be taken to the hospital because he swam in algae-infested water, and also in Genoa, more than two hundred people were sent to hospital after swimming in the algae or inhaling toxins carried to the beach by the wind. During the summer of 2009, officials in Massachusetts put out a toxic-beach-algae warning but did not close the beaches. And the problem extends to freshwater lake beaches as well.

Some are attributing the algae outbreaks to global warming. Although this may indeed be a factor as our seas warm up, it is clear that excess nitrate-rich fertilizers, along with animal wastes and poorly treated or untreated sewage, are the main villains.

The problem is greater than just the hazard it poses to humans. When a beach is covered with algae, virtually everything that lives on and within the beach is killed, while access is denied to nesting and to food for local birds, fish, sea turtles, and various crustaceans. Thus, an entire beach–nearshore ecosystem, which includes microscopic organisms (meiofauna) living between sand grains at the bottom of the food chain up to sharks cruising offshore, is wiped out. Simultaneously, oxygen is usually depleted in nearshore waters, which creates a threat to marine mammals and seabirds.

Although politicians are finally beginning to pay attention to this danger and take action, as in France, the problem comes down to the use of fertilizers to boost food production on farms and agribusinesses along the coasts and nearby rivers. They must recognize their responsibility to be good stewards and adopt agricultural methods that protect the coastal waters. Beaches are not only habitat, but the basis

Gulf of Mexico beaches than one might expect, given the size of the spill. This is not to say that there wasn't beach damage: beaches in Louisiana, Mississippi, and Alabama particularly were affected, as were marshes in Louisiana and the offshore fishery. The true extent of the total damage to habitat may never be known.

Beach cleanup procedures are costly and very damaging environmentally. Sometimes the damage from cleaning a shoreline is probably, in the short run, more damaging than the spill itself. Different shoreline environments have different susceptibilities to spilled oil. A vertical rock cliff would be the least susceptible, followed by more susceptible sandy beaches, tidal flats, and, most vulnerable of all, marshes and mangrove swamps. Fine-sand beaches are less susceptible than coarser-sand beaches because in

fine sand the oil penetrates only about 0.5 inch (1.3 cm) due to lower permeability. The coarser the sand on a beach, the more damage a spill will cause because of increased penetration. Gravel and boulder beaches are particularly vulnerable (e.g., as in the impact of the *Exxon Valdez* spill). Thus, high-latitude beaches on glaciated coasts are particularly susceptible. Arctic beaches are commonly made up of gravel, and many tropical beaches associated with coral reefs are very coarse sand and gravel; hence, all are susceptible to spills.

Bubbly sand (see chapter 7) and the effects of burrowers (see chapter 9) counteract the grain-size effect. Oil has been observed to penetrate into the beach as much as 1.3 ft (40 cm) in bubbly sand and to greater depths in deep

Coauthor Joe Kelley is leaning on a septic tank exposed on the beach after a nor'easter caused extensive shoreline retreat at South Nags Head, North Carolina (November 2009). The septic tank had been exposed in previous storms and was then reburied with trucked-in sand. After this storm, however, the houses were condemned and the owners were ordered to remove them. In the background, beyond the line of light-brown sandbags, is a blue house that was in the third row of houses back from the beach twenty-five years ago. Now the house is in row number one, next to the beach, waiting its turn to fall into the sea.

Upper right Swash lines made up of spilled oil (69 million gallons) from the 1978 *Amoco Cadiz* disaster off the coast of France. Oil penetration is not as severe on fine-sand beaches providing the sand is well sorted, not bubbly, and not burrowed; however, not many beaches have this set of properties. Photo courtesy of Jacqueline Michel.

Lower This dark pool of oil on Isle Grand Terre, Louisiana, from the May 2010 BP Deepwater Horizon oil spill is only a small local example of the extensive environmental disaster that spread along the Gulf of Mexico coast. Every high tide brought more oil to the beaches, sounds, and marshes. The penetration of the oil into the sediment is a function of the grain size: The coarser the grains, the greater the penetration is the rule. Needless to say, all living beach creatures will probably be killed in a locality such as this. Note the drilling platform in the far background. As these facilities age, and as drilling proceeds into deeper water and less stable seafloors, more such spills will occur. Photo courtesy of Adam Griffith.

Upper left An oil-covered boulder beach along the shores of Prince William Sound in Alaska, from the 1989 *Exxon Valdez* oil spill (11 million gallons). Oil penetrates deeply into boulder beaches and is almost impossible to remove, as attested by the fact that oil from this 1989 spill is still present in beaches more than twenty years later. The process of oil removal is always environmentally damaging in and of itself. Photo courtesy of Jacqueline Michel.

burrows associated with crabs and ghost shrimp. Thus, on this account, the vulnerability of beaches on the southeastern U.S. Atlantic will vary considerably. Georgia beaches have both more *Callianassa* burrows and more and thicker bubbly sands than beaches in North Carolina and therefore are more endangered by oil spills, even though they are finer grained than the North Carolina variety.

Oil spilled on beaches eventually disappears, but how quickly it goes away depends on the amount and type of oil that was spilled and the local climate. Early in World War II, many beaches in Europe, the Caribbean, and along the U.S. East Coast were blackened by spilled oil from tankers sunk by submarines. Most beaches recovered in a few years.

An oil-covered beach in Urquiola, La Coruña, Spain, in the "cleanup" after a 1976 oil spill, clearly a horror story. Photo courtesy of Miles Hayes.

THE ENVIRONMENTAL TRUTHS ABOUT BEACHES

Beaches have found the fountain of youth. Absent the impact of humans, few natural beaches ever need salvation.

Beaches protect themselves from storms. Beaches flatten and produce offshore bars during storms, processes that reduce the impact of big storm waves and limit beach retreat.

Beaches have different personalities. Different types of waves, the frequency of storms, and variability in grain size, composition, and the sand supply give each beach its uniqueness.

Beaches never create an erosion problem. Those who build next to the beach create the problem.

Beaches are damaged by shoreline engineering. Whether by the emplacement of an artificial beach or by the construction of a

Beach mining goes on at every scale, from the local horse-and-cart operation to the giant mineral mines in coastal sands, such as in Namibia. Here a young man is obtaining sand from a beach in Uruguay near Cabo Polonio. When you multiply the volume of this individual's sand removal by many thousands, it can become a significant factor in beach degradation. Photo courtesy of Rob Thieler.

boulder seawall, engineering alters beach dynamics so the natural beach and its ecosystem will never be the same.

The greatest single threat to beaches in the immediate future is engineering. The extensive global coastal engineering effort is largely to save buildings, not to save beaches.

The greatest single threat to beaches in the longer term (more than fifty years) is engineering. The initial response to retreating beaches due to rising seas has been and will continue to be attempts to hold the shoreline in place with engineering structures and nourishment sand. The resulting costs, both economically and environmentally, will increase exponentially as it becomes more and more difficult to defeat nature over the great length of developed shorelines.

Globally, beaches can be viewed as endangered habitats. Unless there is a significant change in the philosophy of conserving the global shoreline, natural beaches will continue to be replaced by artificial beaches and become extinct, just like endangered species. Both current and future management of beaches will require much care, improvement, and foresight if the beaches are to survive in the latter half of the twenty-first century and beyond.

13

THE URBANIZED BEACH: FROM MIDDENS TO THE MAELSTROM OF DEVELOPMENT

Many modern beaches can no longer be seen in terms of the natural categories outlined in the chapters of this book; urbanization has created a new category of shoreline, sometimes beachless, sometimes characterized by a faux beach. Urbanized beaches are the product of a historical pattern, or evolution, from subsistence beaches to resorts to complete urbanization.

People originally used beaches as places to hunt birds, fish, or launch boats (i.e., for fowling, fishing and navigation). In warm climates, people certainly cooled off along beaches, but the most common record we find of the ancient human occupation of beaches is in shell middens. *Middens* are prehistoric garbage dumps, usually composed of discarded shells from the shellfish that inhabited adjacent tidal flats and marshes. People had little impact on beaches in prehistoric times.

Swimming and bathing were generally not practiced by Christians in the European Middle Ages in part because of the association of those practices with Roman decadence and in part because of theories that immersion made the body susceptible to disease. Northern Europeans maintained only a practical use for beaches until the 1700s, when the Grand Tour brought wealthy, educated people to the Mediterranean (e.g., the French Riviera and the Italian coast) and elsewhere, where they observed beautiful and sometimes desolate beaches. Painters created romantic images of exotic beaches as well as of rural beaches with local fishermen. The wealthy traveler sketched beach scenes and wrote lyrical prose about the wonders of the seashore. By the mid-eighteenth century, the British popularized bathing in the sea more as a health measure

(Captions on facing page)

Upper left Bagdad Beach, Mexico, is an example of a low-cost beach resort, sited at the high-tide line of a very flat beach, which is migrating under the buildings. This beach is immediately south of the U.S.-Mexican border on the Gulf of Mexico and is used mainly by Mexican citizens.

Upper right Everybody enjoys the beach, but in a variety of different ways. Here the ladies' circle is engaged in a competitive card game, next to a seawall in Cadiz, Spain.

Center left A mobile refreshment stand on Noosa Beach, Australia, provides an example of how to make sure that the concessions are where the people are on a given day at the beach. The need for permanent buildings at the back of this eroding shoreline (note the scarp in the background) is eliminated.

Center right Typical beach chairs for hire on the beach at Egmond, Netherlands, provide shelter from cool breezes (and perhaps blowing sand) while the occupants enjoy the summer sun.

Lower left A draft-horse–drawn wagon carrying tourists around on Texel Island, Netherlands.

Lower right This tram incline in Bournemouth, England, provides an unusual type of beach access for beach users to get past the steep cliff.

than a recreational matter, and seaside spas grew in scale and number. These were designed for the wealthy, and sea bathing was accompanied by walks along the strand and fine dining during the summer season. The beaches at Brighton, Margate, and Scarborough, England, drew large numbers of visitors in the mid-to-late eighteenth century and were among the first seaside resorts. In France as well, fishing villages in Biarritz and the Riviera became transformed into resorts for the wealthy, and other resorts sprung up around Europe's coastline. Seaside resorts were made widely popular by poets, impressionist painters, and later, photographers of the nineteenth century. Rail and ferry services grew to efficiently bring larger numbers of tourists to the developing seashores, and hotels and restaurants replaced the fishing villages. Boardwalks separated the grand hotels from what remained of sand dunes. Following storms, little by little, seawalls were built on the outer edge of the boardwalks to protect property. Many coastal cities are now fronted by massive esplanades—the combined concrete and stone seawalls, promenades, and avenues. Horse-racing and gambling establishments were built to entertain the crowds, who no longer came to the beach for their health.

Beach resorts in the United States began in Cape May, New Jersey, and spread northward up the coast toward New York City. American resorts were patterned after later-model European resorts and were intended more for the masses who came by train than for the wealthy. There were four hundred hotels in Atlantic City, New Jersey, by 1900, and brothels, drinking establishments, piers that extended out into the surf, and other nonbeach forms of entertainment dominated coastal resorts. At about this time, the first large amusement park, including a Ferris wheel, was established at Coney Island, New York (a barrier island).

After World War I, the French and Italian rivieras grew as major resorts for the sun-bathing wealthy of Europe. In the United States, Florida was discovered and rapidly developed. Palm Beach, Florida, served the very wealthy, and Miami Beach appealed to the middle classes seeking sun and sand. World War II brought soldiers to the beaches

This hotel in Waikiki has lost most of its beach in front of its seawall. In order to provide a "beach" for its customers, the hotel has constructed a concrete box and filled it with sand that is connected to the beach. Some people called it the "sand-box," a sad commentary on the state of beaches along heavily developed shore-lines. Photo courtesy of Norma Longo.

This rock revetment along the shoreline of Kalama Park on Maui, Hawaii, prevents erosion, but at the cost of the beach. Sand may have been taken from this shoreline, which created the problem, but if erosion had been allowed to continue, a new beach would have formed. Photo courtesy of Norma Longo.

Closely spaced lounge chairs facing the beach at Cannes, France, one of the world's premier beach resorts, reflect the end result of recreational beaches narrowing between the sea and the esplanade seawall. Space per person is limited, and the beach is no longer the open expanse that many associate with beaches. Yet, its appeal as a social gathering place remains for many.

of Europe and the Pacific, and the graveyards, monuments, and fortification ruins still attest to the world's brutal use of beaches in the 1940s.

After World War II, the recreational development of coastlines took on modern appearances. Air travel put all the world's beaches within reach of those who could afford to fly, and cars brought the rest to the coast. Trailers and growing numbers of second homes created a beach lifestyle that many desired. Air-conditioning transformed the indoors of tropical resorts into tolerable climates, while the outside remained sunny and hot. By the 1960s, beaches had become a middle-class cultural focus for young people in the United States. Beaches were now for showing off fast cars, surfing, and weight lifting, and films and music attempted to document this new lifestyle.

Every coastal resort has a different story, a different beginning. Australia's Gold Coast was a lightly settled area beginning in the mid-1800s, but with the construction of the Surfers Paradise Hotel in 1925 the tourist rush began. In the 1970s, developers began to rule the roost, and construction of high-rise buildings began. Air service to the region began in 1981.

Post–World War II resort development increased in many of the locales that servicemen and women discovered in their travels during the war. Florida and California in the United States, southern Europe, and notably the Hawaiian Islands are good examples.

In the 1950s, Spain's Costa Brava region of northeastern Catalonia was designated by the Franco government as a holiday destination. Thus began the rush to the sea, and tourism quickly replaced fishing as the principal local industry. It began as a package holiday destination for tourists from France and the United Kingdom and gradually broadened its appeal to a wide variety of vacationers. The history of Portugal's Algarve coast is similar; it remains a popular destination for northern Europeans.

The Côte d'Azur, or French Riviera, was one of the first modern beach resorts, beginning in the late eighteenth century as a health resort for wealthy British and quickly evolving into a favorite beach resort for the aristocracy and wealthy of all of Europe. Before World War II, it was a gathering spot for the wealthy of many nationalities and for artists and writers such as Pablo Picasso, Edith Wharton, and Aldous Huxley. After the war, it remained a popular tourist destination, but no longer just for the upper class.

Only three people lived on the Isla Cancún in 1970 when the Banco de Mexico began hotel construction on Cancún Beach. Construction followed a lengthy period of research to find the best location for Mexico's newest and some say greatest beach resort. Today there are 150 hotels with 24,000 rooms and 380 restaurants in the resort development, and the popularity has led to extending the development along the coast to the south, the Mayan Riviera.

All good things end, however, and resorts come and go, in and out of fashion as the crowds of early visitors transform the natural attractions of the seaside into an artificial, crowded, polluted landscape often beset with crime. The stages in the life and death of

Is this the view of the future? A child sits alone where once a dune stood, and still her beach is being removed—not by nature's waves and wind but by development's thirst for sand. Photo courtesy of Lana Wong.

a beach resort are directly related to the number of tourists a location draws. When a future resort was first discovered, it was often a fishing community (e.g., the coastal resorts of Portugal and Spain). Visitors came and local people adapted to accommodate the visitors by opening bed-and-breakfasts, operating boat rides to barrier islands, or taking the more adventurous on fishing expeditions. Rarely can one find a truly undiscovered beach today, although they exist in some remote parts of South America, Africa, Southeast Asia, some islands, and around the Arctic and Antarctic. Over time, more visitors came and hotels were built near entrance points, at first near rail connections in Europe or along new beachfront roads in the United States. Once development took off, all locations along the seaside were built upon. No longer were local people needed; the money and staff to run the resorts came from distant cities, and the profits flowed back there. To protect the investments from natural shoreline-change processes, such as storms, seawalls and groins were erected. These also had the undesired effect of causing beach loss, but visitors now came for more than the shore. Gambling and other adult-oriented activities replaced seaside bathing and family recreation. The numbers of visitors inevitably declined as the beach resort became seedy and natural disasters more difficult to survive. Last Island (Isle Dernière), Louisiana had a hotel

A view of Honolulu and Waikiki Beach, one of the world's great beach resorts. The actual beach here is very narrow and exists only through repeated artificial nourishment. Note the small pink building on the shore in the middle of the row of high-rise buildings. This was the original "large" hotel on the shoreline, the "seed" that led to this field of tourist hotels. The future of Waikiki and other beaches like it will depend heavily on the rate of future sea-level rise and the ability to find an affordable sand supply for continued nourishment. Certainly by the end of the twenty-first century, with the expected minimum sea-level rise of 3 ft (almost 1 m), the cost of beach nourishment to maintain this primary tourist attraction will be extremely high. Photo courtesy of Norma Longo.

with a thousand rooms in 1856, but it was so devastated by a hurricane and associated loss of life that it was never rebuilt. Water was polluted by inadequate sewage treatment and fish disappeared. In this stage of stagnation and decline, a beach resort lacks the means to deal with storms and fires, so large areas of old resorts were often abandoned.

Many resorts came back after storms or fires eliminated much of the unsightly development. The new development, although sometimes more in harmony with nature, often was touted as "bigger and better," which meant more valuable property at greater risk. Sometimes casino gambling led to beach-resort rejuvenation, as at Atlantic City, New Jersey; at other times beach replenishment made the seaside attractive again, as at Miami Beach, Florida. Thus, most of the developed world's beaches present an appearance today that is very different from that of the past. The oldest resorts, like Cape May, New Jersey, and Brighton, England, have lost much of their original appeal (and their beaches as well) as erosion and neglect have haunted them.

Some attractive beaches avoid the resort model of development and remain attractive. These are found in national parks and seashores, where management generally means letting natural processes, including erosion, happen. On the beaches of Cape Cod, in Massachusetts, for example, "The Outermost House" of literary fame was allowed to succumb to the sea, as were the signal towers of Marconi's original wireless radio device. When a storm eroded a bathhouse and parking lot, the rubble became buried under a modern sand dune, and nothing was rebuilt. Exceptionally valuable property, like Highland Light, established by George Washington, was moved to prevent its destruction. In these locations, the number of visitors is sustained at relatively high levels for a long time even as parkland is lost.

Some wealthy residential complexes are also relatively natural. The cost of land has limited development, and in some instances prudent placement of properties back from the beach has minimized the need for shoreline armoring. Despite the best intentions of maintaining a beach as a developed paradise, in places like North Topsail Beach, North Carolina, loss of developed beach properties is causing residents to consider engineering their shoreline.

Beach resorts for future generations will be different from today's beach playgrounds. The anticipated sea-level rise by 2100 will make maintenance of beaches through the process of placing sand on the beach untenable. Many beach resorts two generations from now will likely be heavily seawalled and beachless. Older beach resorts will follow the example of Cape May, New Jersey, where the principal activity will be promenading on top of a seawall on a beachless shoreline, watching the waves break and taking the sea air. Developers will seek unspoiled areas where beach retreat has been unfettered by buildings and will repeat the same history as for past resorts. Only in parks and national seashores will natural beaches remain to be enjoyed, as the most dynamic systems on Earth continue to move landward.

GLOSSARY

ACCRETION The opposite of erosion. On a beach, sediment accumulation results in beach widening and seaward growth.

ADHESION RIPPLES A feature formed when dry sand blown over a wet beach adheres to the damp surface, to form an irregular, warty bed form.

AEOLIANITE See *eolianite*.

ANTIDUNES Small, current-formed, asymmetric ripple-like features roughly parallel to the shoreline that develop in the swash zone, only inches high, with their steep face oriented opposite the direction of backflow. They form on beach surfaces where rapid seaward backwash returns over fine-grained sand. Antidunes are usually washed out on the falling tide but leave a distinctive striped pattern on the beach. Somewhat larger antidunes also form in channels where the flow velocity is high.

ARMORED MUD BALL A piece of mud that has rolled on the beach to become coated with sand and coarser material, often shell fragments, which "armor" the mud, slowing the ball's attrition.

BACKSHORE That part of the beach landward of the normal high-tide line.

BACKWASH The seaward flow of water return from wave swash, down the beach face.

BARRIER ISLAND A narrow sandy island, usually close to and paralleling the coast, but facing the open ocean and bounded on each end by tidal inlets. Barrier islands make up about 12 percent of the world's open-ocean shoreline and are characteristic of trailing-edge coasts (e.g., the Frisian Islands of the Dutch-German North Sea Coast and the East and Gulf Coasts of the United States).

BARRIER ISLAND MIGRATION The landward movement of an island in response to either a rising sea level (Colombia's Pacific Coast) or a loss of sand supply (Nile Delta).

BEACH BISCUITS An informal name for biscuit-shaped lumps of cohesive sand on the upper beach surface, often occurring in a regularly spaced pattern. In some instances, these are the resistant caps that lead to the eventual formation of pedestals.

BEACH GLASS Collectors' informal name for rounded pieces of broken glass found on beaches. Also called mermaid's tears and lucky tears.

BEACH GROOMING Any cleaning of a beach by raking, sweeping, or mechanical sifting to remove wrack and debris.

BEACH NOURISHMENT The artificial rebuilding of a beach by the addition of sand from an outside source to hold the shoreline in place. Also called beach replenishment, and dredge and fill.

BEACH PROFILE A line representing the silhouette of the beach as viewed when looking parallel to the shoreline.

BEACH RIDGE An elevated berm above normal high tide, typical of a beach that is accreting or building seaward. The term is also used to describe the first row of dunes landward of the back beach.

BEACH ROCK Cemented beach sand that forms ledges between the high- and low-tide lines.

BEACH SCRAPING / BEACH BULLDOZING A procedure in which a layer of sand is bulldozed or pushed from the intertidal zone and forebeach to the back of the beach in order to form an artificial dune or sand dike. This is usually done to protect buildings.

BEDFORM A small-scale feature formed on the surface of the beach or dune. Also called sedimentary structures. Examples include lineations, rills, and ripples.

BERM A depositional terrace-like feature on the upper beach.

BERM CREST The seaward edge of the berm, beyond which is the steep berm face, sloping seaward.

BIOGENIC A term that refers to sediment produced by organisms.

BIOTURBATION Sedimentary structures (seen in cross section or in a ditch) disturbed or destroyed by the activity of animals and plants (e.g., plant roots or crab burrows).

BLACK SAND See *heavy minerals.*

BLISTER A small bedform on the beach resulting from trapped air pushing up a miniature mound of sand, usually circular or oval in plan view. Characterized by an internal air cavity.

BLOWOUT A bowl-shaped or flat area in a dune field where the sand has been blown away (eroded). This erosion usually proceeds until the groundwater table is reached, at which point the wet sand hinders further sand removal. This results in a flat blowout floor. This is the only large-scale erosional feature formed by winds.

BREAKWATER Hard structure placed offshore, usually parallel to the shore, designed to cause waves to break, lessening erosion on the beach immediately behind the breakwater. Breakwaters interrupt longshore drift and cut off sand supply to downdrift beaches.

BUBBLE TRAINS Lines of bubbles formed by air escaping from a hole in the beach into the water layer of the swash or backwash. The bubble line may be straight or curved depending on how the swash is moving across the beach.

BUBBLY SAND Sand with cavities or very large visible pores that are formed when air bubbles are trapped in the sand. See *soft sand*.

BULKHEAD A relatively low and small seawall designed not to protect buildings from waves, but to keep land from eroding from behind wall.

CALCAREOUS A term used to describe material composed of calcium carbonate ($CaCO_3$). On beaches this is usually shells, skeletal material such as coral, and the sand derived from these sources.

CARBONATE FRACTION The calcareous portion of the sediment composed of grains of calcium carbonate ($CaCO_3$).

CHIMNEY STRUCTURE The resistant rim and burrow-tube lining that protrudes out on a beach surface when exposed by swash erosion.

COASTAL PLAIN A broad, gently seaward-sloping plain, often tens to hundreds of miles wide and underlain by sedimentary rocks, that borders the coast. Coastal plains occur primarily on trailing-edge coasts (e.g., eastern North and South America and East Africa).

COLLISION OR LEADING-EDGE COAST A coast that borders the colliding oceanic and continental plates. Usually the continental margins here are narrow and steep and bordered by mountain ranges, and the beaches tend to have relatively high waves (e.g., western North and South America).

CONTINENTAL SHELF The very gently sloping surface extending from the beach out to where the steeper continental slope begins, usually at depths greater than 300 ft (90 m). The width of the continental shelf is an important control of the size of the waves that strike the beach; narrow shelves (Peru, Chile, eastern Japan) favor high waves, and wide shelves (Argentina, western Australia, eastern China) favor low waves.

COQUINA Calcareous rock, composed mostly of shell material, cemented by calcium carbonate.

CRESCENT MARK A scour mark around any small object on the beach, such as a shell or pebble, that results from wind or water flow around the object. Also called an obstacle mark.

CROSS BEDDING Inclined laminations and beds that form in a sand dune. Successive sets of laminations show different orientations.

CROSS RIPPLE See *interference ripple*.

CURRENT RIPPLE MARK A type of asymmetric ripple mark that forms by water current or wind flow on the beach or dune surface. The steep face of the ripple mark faces in the direction of current flow.

CUSPATE FORELAND A triangular coastal landform (e.g., a cape) resulting from deposition of sediment transported by longshore currents, usually from two directions, and fringed by beaches. Called a ness in Great Britain.

CUSPS Regularly spaced (3 to 9 ft [1 to 3 m]) small embayments (of variable size), separated by small ridges or horns, along a beach that give the shoreline a sinuous or cuspate appearance. The spacing can be considerably greater depending on the size of the waves that formed these features.

DISSIPATIVE BEACH Above the low-tide line these are low, flat, and narrow beaches with a wide surf zone (often more than 325 ft [100 m] wide) on the broad underwater portion of the beach. Offshore sandbars are usually well-developed. Waves are characteristically high, and because of the gentle slope of the offshore beach, the waves lose much of their energy as they first break well offshore, re-form, and break again close to the exposed beach.

DOWNDRIFT The dominant direction of the longshore current flow; analogous to *downstream* in rivers.

DOWNWELLING The flow of surface water into deeper water. This occurs when onshore winds pile up water on the beach, which then escapes by flowing down the surface of the shoreface.

DRAG MARKS Marks on a beach surface resulting from objects being dragged over the sand by waves or swash, scouring the surface. The object or tool may be a shell, seaweed, foam, or other flotsam, and each object leaves a unique mark.

DREDGING / DREDGE AND FILL The removal of sand by dredges to provide sand to nourish a beach.

DRIFT LINE An accumulation of natural and artificial debris (e.g., seaweed, driftwood, plastic refuse, bottles, fishing net floats, lumber) at the most landward extent of the wave swash. See *wrack*.

DRY BEACH See *wet-dry beach line*.

DUNE A feature formed on land from the accumulation of windblown sand, either bare or covered with vegetation. Internally cross bedded due to sand deposition on the various slope faces of the dune.

DUNE CROSS BEDDING A feature in dunes created by the wind building consecutive layers of sand on top of each other that are inclined at different angles. Each single layer indicates a former surface of the dune.

DYNAMIC EQUILIBRIUM The natural balance between processes and materials in the formation of any landform, including beaches. Sea level, wave energy, and sediment supply will produce a specific equilibrium beach profile. A change in any factor will cause the beach profile or beach location to change, moving to a new equilibrium.

EBB CURRENT A tidal current formed when the tide is falling, or "going out."

EBB TIDAL DELTA The body of sand that protrudes seaward of an inlet formed by sand carried by ebb tidal currents and shaped by waves.

EMBRYONIC DUNE A small sand dune formed by the initial process of dune formation when a sand deposit forms in the wind shadow of a large obstacle or around a clump of vegetation.

EOLIANITE Windblown dune sand cemented into rock, usually a limestone as a result of dissolution of some calcium carbonate grains and then re-precipitation as cement. Primarily found in the tropics.

EROSION A term used to describe shoreline retreat. The opposite of accretion.

ESTUARY The mouth of a river valley flooded by the rising sea where fresh water and salt water mix.

FETCH The distance of open water over which the wind can blow to form waves. The greater this fetch, the larger the potential wave height.

FLASER RIPPLE MARKS Ripple marks with thin layers of mud in the trough between the crests.

FLAT-TOPPED RIPPLE MARK A ripple mark resulting from a reversal of flow that truncates the crest of a previously formed ripple, leaving a flat top; common on tidal flats where the tidal currents reverse direction.

FLOATING SAND Patches of sand grains floating on a calm water surface, as on standing water in a runnel or at the uppermost fringes of the swash. The sand grains float as a result of the normal surface tension of water.

FLOOD CURRENT A tidal current formed as the tide is rising.

FLOOD TIDAL DELTA The body of sand deposited landward of an inlet, formed by flood-tidal currents.

FOREDUNE The first dune line at the back of a beach.

FULGURITE A rare tubular structure of natural glass found in sand dunes and at the back of dry beaches; formed when lightning struck the sand, causing it to melt and fuse.

GRADED BED A bed that is coarser at the base and becomes finer toward the top as viewed in cross section. The opposite pattern is reverse grading.

GROIN (GROYNE) A hard erosion-control structure installed perpendicular to the beach, designed to trap sand traveling with the longshore current. Shorter than jetties, groins are usually placed in groups (or fields). Groins cause sand accretion on the updrift side but erosion on the downdrift side.

GROUNDWATER Water stored underground in the fractures or in pore spaces between the grains of sand. In a beach, the groundwater may be fresh from overland runoff, or salty from wave and tide waters soaking into the beach.

GROUNDWATER TABLE The top of the zone of saturation where groundwater completely fills the pores. On beaches where the water table intersects the surface, seeps or springs may emerge at low tide.

HARD STABILIZATION Any hard engineering structure designed to trap sand or dissipate wave energy in order to hold the shoreline in place, usually to protect buildings.

HARROW STRUCTURES Linear sand deposits that form in the lee or wind shadow of any obstacle. Miniature harrow structures occur on damp surfaces where larger particles form similar obstacles to fine-sand movement.

HEAVY MINERALS Minerals that weigh more than quartz and feldspar (the light minerals). Commonly these are found in patches of black sand that form during storms. They become concentrated by the winnowing effect of wind and water (placering). Most concentrations of heavy minerals in beach sand are black, often due to the abundance of black grains of magnetite. Sometimes heavy minerals impart other colors to beach sand, such as green sand from olivine (Hawaii) or reddish brown sand from garnet (Labrador).

HOLOCENE A unit (Epoch) of geologic time extending from about ten thousand years ago to the present.

IMBRICATION A pattern of overlap stacking of pebbles or cobbles (clasts), common where these pebbles have a flat shape. This stacking results from waves or currents pushing the clast over in the direction of flow, so the resulting orientation of individual pebbles is sloping down in the direction from which the wave or current came. Shingle beaches usually show imbrication.

INLET A narrow waterway between two barrier islands or barrier spits that connects the sea and the lagoon and is maintained by tidal currents.

INNER BAR The landward bar where two or more submerged sandbars occur off a beach.

INTERFERENCE RIPPLE A compound ripple mark resulting from one set of ripples superposed on an earlier set, and produced by waves or currents coming from more than one direction. See also *ladderback ripple*. Found in runnels and on tidal flats.

INTERTIDAL ZONE The area of the beach between the high-tide line and the low-tide line. The wet portion of the beach exposed at low tide.

JETTY A long, shore-perpendicular engineering structure, often placed in pairs on the sides of inlets to stabilize a navigational channel and to prevent longshore-transported sediment from clogging the channel.

LADDERBACK RIPPLE A compound ripple mark produced when one set of wave ripple marks forms atop of a previous set at nearly a right angle. The resulting pattern looks like a stepladder or is a checkered pattern. Usually produced by changing wave directions during the falling tide, and commonly found on tidal flats.

LAG DEPOSIT A residual layer of coarser sediment that forms on a beach surface when finer sediment is preferentially removed.

LAGOON The body of water between a barrier island and the mainland; the area in which salt marshes develop.

LAMINAE Very thin layers of sediment, usually deposited by a short-term event such as an individual wave or a burst of wind. A lamination, or lamina, may be no more than the thickness of a single layer of sand grains.

LITTORAL DRIFT The sediment transported along a shore by longshore currents. The "net littoral drift" is the volume of sand moved in one direction minus the volume moved in the other direction over a year's time.

LONGSHORE CURRENT A surf- and breaker-zone current that flows parallel to a beach, created by waves striking the shore at an angle. This is the current that transports sand and carries swimmers away from their beach towel.

MACROFAUNA Animals large enough to be visible to the naked eye.

MANGROVE SWAMP A tropical to subtropical marine wetland that is dominated by mangrove trees and bushes; similar in position and biological role to salt marshes which grow in cooler waters.

MEGARIPPLES Large ripple marks with crest spacing between 2 and 20 ft (0.6 and 6 m) deposited by strong currents such as on tidal deltas.

MEIOFAUNA Tiny animals that live between the sand grains of the beach.

MERMAID'S TEARS See *beach glass*.

MICROFAUNA Microscopic animals.

MIDDEN An accumulation of waste material from a cultural site occupied by a group of people (e.g., a fishing or hunting camp) over a long period. In coastal areas, middens usually consist of shell heaps from shellfishing cultures. Also called kitchen midden.

MINERAL SUITE The group of minerals that make up the heavy-mineral fraction of a particular beach. The minerals reflect the composition of their source rocks, as well as their history of weathering and transport.

MUD BALL A ball-shaped clump of clay or peat of variable size produced by the erosion of clay or peat outcrops on the beach. Sometimes mud balls result from beach nourishment using muddy sand.

MUD CRACKS A pattern of cracks or lines formed on the surface of a mud layer due to its drying out and shrinking (desiccating).

NAIL HOLE An informal term for holes in a beach produced by air escape or truncation of air cavities in the beach; so named because of their range of sizes similar to nails. Also called sand holes.

NEAP TIDE The minimum tidal range at a beach (least difference between high and low tides), occurring during the first or third quarters of the moon.

NESS See *cuspate foreland.*

NOURISHED BEACH An artificial beach that has been maintained or widened by the import and addition of new sediment.

OBSTACLE MARK See *crescent mark.*

OFFSHORE BAR An underwater sand ridge offshore from the beach, often identifiable by a line of waves breaking on the bar.

OOIDS Sand-size spherical grains formed by the precipitation of calcium carbonate in concentric layers around some tiny nucleus. The grains usually form in shallow, wave-agitated marine waters. Also called oolites.

OSCILLATION RIPPLE MARK A ripple with a symmetric cross section, produced by the to-and-fro motion of waves. Characterized by a sharp, straight crest and rounded trough. Also called wave ripple.

OUTER BAR The outermost sandbar where two or more bars occur off a beach. The biggest waves break on this seaward-most bar.

OVERWASH Beach sand that has been transported inland beyond the beach by storm waves or the process of waves crossing the beach and moving into the dunes.

OVERWASH FAN See *washover fan.*

PARABOLIC DUNE A U-shaped dune, often formed where a breach or blowout occurs in a foredune in coastal areas.

PARTING LINEATION Faint linear patterns on beach surfaces produced by the swash orientation of the grains. Sand grains usually have at least one axis longer than the other two, so individual grains can be aligned preferentially by the swash, resulting in parallel streaks; like brush marks.

PEAT BALL See *mud ball.*

PEDESTAL STRUCTURE A toadstool-shaped feature formed by differential wind erosion of damp and dry sand layers, or by removal of sand from around an obstacle, leaving it standing atop a small column of sand.

PELLETS Fecal material from organisms of various kinds, often cylindrical in shape and often found on a beach surface surrounding burrows.

PIT Any small depression in the beach formed by a variety of processes including collapse (e.g., blister or air-bubble collapse) and impacts (e.g., raindrop impressions).

PLANE BED A sand layer that has accumulated in a parallel, planar fashion to produce a bed of sediment.

PLANT ARC An arc-shaped ring or other traced pattern in the sand produced by the wind movement of the end of a plant frond, leaf, or rootlet. See *scribe marks*.

PLATE TECTONICS The theory that the uppermost crust of the earth is divided into plates that move about and interact with one another.

PLEISTOCENE A unit (Epoch) of geologic time that extends from about 2 million years ago to the beginning of the Holocene Epoch (ten thousand years ago); informally called the Ice Age.

PLUNGING BREAKER A type of breaking wave that forms on a moderate beach slope (usually 3 to 11 degrees). The breaker curls over, forming a barrel or tube of air as it collapses. This is the most forceful type of breaker in terms of generating sand movement on the seafloor and also the one that surfers dream about.

POCKET BEACH A beach confined between two headlands.

POSITIVE INTERFERENCE When two (or more) waves from different sources meet in phase (crest to crest or trough to trough), they are additive. The resulting wave is the combined height of the two original waves. On the open ocean, this phenomenon can result in very large rogue waves.

PSAMMON General term for all of the tiny plants and animals that live in the beach or move through the sand.

REFLECTIVE BEACH A steep beach without offshore bars, subjected to relatively low waves. The waves travel right up to the beach and break on it. The intertidal zone is narrow, but often the dry beach is wide. Reflective beaches form a continuum of intermediate forms with dissipative beaches.

REVETMENT A common type of seawall built directly on a surface such as the seaward slope of a dune. Revetments are frequently constructed of boulders; the large rocks provide ample interstitial cavities that absorb some of the water from a breaking wave, thereby reducing sand-removing wave reflection and backwash.

RHOMBOID RIPPLE MARK Diamond-shaped ripples generated by backwash over the beach surface. Most common on the falling tide. The apex of the resulting V pattern points landward, or upslope.

RIDGE AND RUNNEL The ridges of a ridge-and-runnel system are sandbars on the beach surface. Runnels are the troughs between the ridges on the beach. See *trough*.

RILL MARKS The small erosional channels in the beach face, carved out by either fresh- or saltwater draining out of the beach at low tide. Seaward ends of rills have small deltas (micro deltas).

RINGS A ring pattern on the beach surface produced when a blister or arched sand layer is truncated. Rings are more pronounced where laminae of different colors (heavy minerals) or grain size occur in the beach.

RIP CURRENT A narrow, fast-moving water current that flows seaward from the beach through the surf zone. A major hazard for swimmers.

RIPPLE MARKS Small-scale ridges separated by troughs or depressions in the sand, typically in a repetitive pattern, and usually less than 2 in (5 cm) in amplitude. These bedforms occur in a wide variety of shapes and patterns depending on air, water, and wave conditions.

RIPPLE TRAIN A set of current ripples moving together.

ROGUE WAVE A rare wave that forms when waves coming from more than one direction converge in phase to create a wave with a very large amplitude. It is believed that some rogue waves have struck beaches.

RUNNEL See *trough*.

SALCRETE Thin sand layers that form fragile crusts, weakly cemented by salt from seawater; they are slightly more resistant or cohesive than underlying dry sand layers.

SALT MARSH The vegetated wetland of the upper intertidal zone in sheltered coastal areas such as lagoons and estuaries, and characterized by salt-tolerant grasses (e.g., *Spartina* sp.), herbs, and shrubs. Such marshes are usually dissected by tidal creeks.

SALT PRUNING The characteristic seaward-sloping profile of a coastal near-beach vegetation line due to the negative impact of windblown salt spray that prevents leaf formation.

SALTATION A type of sand movement, usually by wind, in which the grains are lifted off of the bed by the current (wind or water) and then fall and strike another grain, launching it into a similar arclike motion.

SAND A size term for grains made up of anything (minerals such as quartz or magnetite, or bits of shells) that range from 1/16 to 2 mm in diameter.

SANDBAR See *offshore bar*.

SAND HOLE See *nail hole*.

SAND VOLCANO A miniature cone structure of sand formed when water escapes from a hole in the beach, either due to the tidal effect or when a burrowing organism forces water out though its burrow opening.

SAND WAVE Submerged depositional features formed by strong currents that have crest-to-crest spacing of greater than 18 ft (5.5 m). They are common where tidal currents flow, as in inlets or over tidal deltas.

SCARP A shore-parallel vertical sand "cliff" as high as 7 feet (2 m) on the beach (berm scarp) or at the toe of the dune (dune scarp), usually indicating rapid erosion (often in storms). Scarps are very common on nourished beaches.

SCRIBE MARKS Marks in the sand surface of a dune or back of beach produced when a plant leaf, frond, or exposed root is moved by the wind against the sand. Often circular or semicircular in form.

SEA Locally generated waves in a chaotic surface pattern that is choppy and irregular (confused).

SEA FOAM Foam that is produced during storms; when the bottom is stirred up, organic matter is released and, when churned by the waves, forms foam.

SEA LEVEL The average elevation of the water surface of the sea.

SEAWALL A general term for any hard structure such as a wall, revetment, or bulkhead installed parallel to the shore on the upper beach in an effort to prevent shoreline retreat and loss of property.

SEDIMENT BUDGET A listing of all the sources of sand that contribute to a beach, including rivers, shoreline erosion, and waves pushing sediment ashore from the continental shelf.

SEDIMENTARY STRUCTURE A general and broadly applied term that includes surface bedforms and internal beach and dune structures, such as cross bedding, produced by sedimentary agents such as waves and currents.

SHADOW DUNE See *embryonic dune*.

SHELL HASH A general term for any concentration of broken shell material on a beach.

SHOALING Becoming shallower, as in a shoaling wave.

SHOREFACE The narrow, relatively steep surface extending seaward from the low-tide line out to the point where the seafloor slope flattens and merges into the continental shelf, often at depths of 30 to 60 ft (9 to 18 m). The shoreface is an extension of the beach on sandy coasts and is the surface of active sand exchange between the inner shelf and the beach.

SINGING SAND Beach or dune sand that produces any sound when subjected to pressure such as walking on the beach, or the wind blowing over a dune. The sound is produced by the shearing of grains against each other and goes by a variety of names including singing, chirping, barking, whistling, and croaking. The low-frequency sound produced by sliding sand on a dune surface is usually described as booming.

SOFT SAND An informal term for beach sand into which one's feet sink deeper than on a firm beach, making it difficult to walk. The softness can be due to the presence of bubbly sand, trapped air in the beach, or noncohesiveness attributable to a combination of factors including amount of water saturation, grain size, and seaweed content.

SOFT STABILIZATION Any approach to holding a shoreline in place through the use of non-hardening techniques, but most commonly the artificial placement of sediment on the beach (e.g., dredge and fill; beach replenishment). Also includes the use of vegetation to stabilize dunes.

SORTING A measure of the range of grain sizes in a sediment. A well-sorted sand has a narrow range of grain sizes and a poorly-sorted sand has a wide range of grain sizes.

SPILLING BREAKER A wave that breaks on a relatively flat beach slope (typically 3 degrees or less). The wave crest spills over the top of the wave but does not curl like a plunging breaker.

SPIT A sandbar or fingerlike beach extending from the land and formed by longshore sediment transport; typically curved or hooklike on its seaward end. Spits extend into inlets and mouths of bays. Spit formation can be an initial stage in the formation of barrier islands.

SPRING TIDE The highest tidal range at a beach, occurring during a full or new moon.

STANDING WAVE A smooth wave that forms and stays in place on a current of high-velocity flow, as seen in beach channels. The sediment on the bottom of the channel has a form similar to the water's surface.

STORM SURGE The superelevation of the water surface due to the combination of low atmospheric pressure at the center of a storm, wind stresses pushing water onshore, and wave setup. Storm surge approaching a coast also is affected by the bottom physiography as water is pushed into shallower depths. Shoreline configuration can also add to the storm-surge effect (e.g., funnel-shaped shorelines are highly susceptible to surges).

SUPRATIDAL FLAT A flat land area landward of the highest tide line, which is occasionally occupied by water from storms. In Abu Dhabi, the supratidal flats are areas of formation of evaporite minerals.

SURF ZONE A band of water within which waves are breaking adjacent to the beach.

SURGING BREAKER A wave that comes ashore on a steep bottom slope (generally greater than 11 degrees) but does not break like spilling or plunging breakers.

SWASH The last remnant of the final breaking wave as it runs up the beach.

SWASH MARK A sinuous line of sand, shell debris, seaweed, or other fine detritus, left at the edge of the uppermost wave swash. A line that marks the maximum advance of a dissipating wave onto the beach.

SWASH ZONE The area of the beach where wave swash and backwash are running up and down the beach slope. The swash zone moves up and down with the tides.

SWELL Regularly spaced waves, with long wavelength and continuous crests, that were formed by winds far from the beach.

TAR BALL Similar to a mud ball but formed from blobs of tar from oil spills or less commonly from natural oil seeps. These are the final remnant of spilled oil left behind as the more volatile fraction evaporates away.

TERRIGENOUS A term describing sediment derived from land.

TIDAL FLAT A more or less flat surface exposed at low tide, sometimes composed of sandy mud. Sandy tidal flats are often covered with extensive fields of ripple marks and the tracks, trails, traces, and burrow openings of animals living on or in the sediment.

TIDE The daily rise and fall of the local water level caused by the gravitational pull of the moon and sun on the ocean's waters as the Earth rotates.

TIDE RANGE The vertical difference between normal high and low tides; synonymous with tidal amplitude.

TOMBOLO A body of sand, usually perpendicular to the shoreline, which connects the beach to an offshore island or rock. Tombolos often connect offshore breakwaters with the beach.

TRAILING-EDGE COAST Coasts usually bordered by coastal plains where the oceanic tectonic plate is moving along with (not colliding with) the tectonic plate that the continent is on. As a rule, the beaches on trailing-edge coasts tend not to be rocky, the continental shelves are wider, and the beaches have sand that is finer and more quartz rich than that of beaches on colliding or leading-edge margins. Examples are the east coasts of North and South America.

TRANSVERSE DUNE See *foredune*.

TROUGH A channel-like feature between berm crests or sandbars (ridges) exposed at low tide; also called a runnel. The trough is usually visible as a drainage channel at low tide and is a locale of current ripples.

TSUNAMI An immense, fast-travelling wave formed as the result of an earthquake, submarine landslide, or underwater volcanic eruption. Tsunami waves are small in the open ocean, but in shallow water, they peak up (increase in height).

UPDRIFT The direction opposite the direction of net longshore drift of sand. Analogous to upstream in rivers, and the opposite of downdrift.

UPWELLING Water moving from deep water to the surface. On beaches this usually occurs when winds are blowing offshore, causing water to upwell to replace water that has moved away from the beach.

VENTIFACTS Rocks that have been abraded, grooved, or otherwise shaped by wind-blown sand that usually comes from a desert floor. The result is similar to sand blasting.

WASHOVER FAN A fan-shaped sand deposit resulting from storm waves that carry sediment beyond the back of the beach into the dunes or backing salt marsh and lagoon. Where adjacent fans merge or coalesce, they form a washover apron.

WATER-SATURATED BEACH A beach in which the pores are completely filled with water and the water table in effect is the surface of the beach. Such beaches often give a mirror-like appearance in the distance, reflecting the sky or skyline.

WATER TABLE The surface of the zone of groundwater saturation. See *groundwater table*.

WAVE The form water takes as energy is transferred from the wind to the sea surface, and consists of the crest (high point) and trough (low point). A wave moves through the water from its wind source area of formation to the coastline. Waves move water in a circular or elliptical rotation, not in a forward direction with the wave form.

WAVE AMPLITUDE Half of the vertical distance between the wave crest and trough.

WAVE BASE The greatest depth at which waves stir up the bottom.

WAVE CLIMATE The characteristic wave conditions (height, period, direction, storm magnitude and frequency) over time for a particular area, and an important factor in determining the type of beach that forms on any given coast. For example, ocean beaches in southern Iceland are usually subjected to high waves, while beaches in Georgia (U.S.) commonly experience very low waves.

WAVE CREST The line formed along the highest point of a wave.

WAVE FREQUENCY The inverse of the wave period, or the fraction of a wave that passes a given point in one second.

WAVE HEIGHT The vertical distance between the wave crest and trough.

WAVELENGTH The horizontal distance between wave crests.

WAVE ORBITAL The internal water movement caused by the passage of a wave. Orbitals are circular in deep water and elliptical in shallow water.

WAVE PERIOD The time it takes for two adjacent wave crests to pass a given point.

WAVE REFRACTION The bending of waves as they come ashore and begin to feel bottom, or change in the direction of the wave front as the wave encounters obstructions such as islands or headlands.

WAVE RIPPLE A type of ripple mark formed by waves and also referred to as a long-crested ripple mark.

WAVE SETUP Water that piles up along a coastline due to continually incoming waves. Water brought in by waves comes in faster than it can drain back to sea, elevating the local water level during storms.

WAVE STEEPNESS The steepness of a wave, measured by the ratio of the wave height to the wavelength.

WAVE TRAIN A group of waves of similar size and speed moving together in the same direction.

WAVE TROUGH The lowest point of a wave.

WET-DRY BEACH LINE The high-tide line marked by the limit of wet sand. The width of the dry beach is one measure of the health of recreational beaches.

WIND RIPPLE MARK A long, parallel-crested ripple mark formed by wind, usually of lower amplitude than water-wave ripples. Typically found on the backshore of the beach and in sand dunes.

WIND SETUP The piling up of water along a coastline due to onshore winds. Winds blow water up against the beach, elevating the local water level, especially during storms, and causing offshore flow of water along the seafloor.

WIND SHADOW The area protected from the wind in the lee of an object or obstacle to the wind, usually resulting in sediment deposition in this position.

WINTER-SUMMER BEACH A highly variable and site-specific characteristic of beaches reflecting the higher waves and more common storms in the winter relative to the summer.

WRACK The general term for debris, whether natural or artificial, that has washed up on a beach, usually accumulating at the edge of the last high-tide line (wrack line) or from the edge of the swash during the last storm. See *drift line*. Flotsam and jetsam from shipping are a common source, but increasingly wrack is dominated by refuse having a land origin.

WRINKLE MARKS Tiny ripple-like structures that form on a mud surface when covered with a thin film of water and subjected to a wind current.

SELECTED REFERENCES FOR
FURTHER READING

The following references are a general guide for those who wish to read more about beaches. An extensive scientific and technical literature exists, and it is not the intent or goal of this book to review or present that literature. Several of the texts cited here provide an introduction to the technical literature, particularly those on geomorphology, coastal dynamics, and sedimentary structures.

Ackerman, J. 1995. *Notes from the Shore*. New York: Viking Press.

Ahmad, E. 1972. *Coastal Geomorphology of India*. New Delhi, India: Orient Longman.

Allen, J. N.d. *Beaches—An Island's Treasure: An Environmental Guide to the Recreational Beaches of the North Irish Coast*. Coleraine, Northern Ireland: Coleraine Borough Council.

Allen, J.R.L. 1982. *Sedimentary Structures: Their Character and Physical Basis*. Vols. 1 and 2. New York: Elsevier.

Alonso, I., and J.A.G. Cooper, eds. 2006. Coastal Geomorphology in Spain. Special issue, *Journal of Coastal Research* 48.

Aminti, P., and E. Pranzini, eds. 1993. *La difesa dei litorali in Italia*. Rome: Edizioni delle Autonomie.

Anderson, J.B. 2007. *The Formation and Future of the Upper Texas Coast*. College Station: Texas A&M University Press.

Anthony, E. 2008. *Shore Processes and Their Palaeoenvironmental Applications*. Amsterdam, Netherlands: Elsevier.

Anthony, E.J. 1990. Environnement, géomorphologie et dynamique sédimentaire des côtes alluviales de la Sierra Leone, Afrique de l'Ouest. *Revue d'analyse spatiale quantitative et appliquée* No. 27–28. Nice, Presses CRDP.

Aono, K. 1980. *Seashores of Japan*. Tokyo: Kyoritsu Shuppan.

Ballantine, T. 1991. *Tideland Treasure*. Columbia: University of South Carolina Press.

Bascom, W.N. 1964. *Waves and Beaches*. New York: Anchor Books.

Baxter, J. 1990. *Pocket Guide to Seashores of Britain and Northern Europe*. London: Mitchell Beazley.

Bird, E.C.F. 1993. *The Coast of Victoria*. Melbourne: Melbourne University Press.

———. 1993. *Submerging Coasts: The Effects of Rising Sea Level on Coastal Environments*. New York: John Wiley and Sons.

———. 2008. *Coastal Geomorphology*. Chichester, UK: John Wiley and Sons.

———. 2008 *Coasts: An Introduction to Coastal Geomorphology*. 2nd ed. Chichester, UK: John Wiley and Sons.

Bird, E.C.F., and M.L. Schwartz. 1985. *The World's Coastlines*. New York: Van Nostrand Reinhold.

Blatt, H., G. Middleton, and R. Murray. 1980. *Origin of Sedimentary Rocks*. Englewood Cliffs, NJ: Prentice Hall.

Britton, J.C., and B. Morton. 1989. *Shore Ecology of the Gulf of Mexico*. Austin: University of Texas Press.

Bush, D.M., O.H. Pilkey, and W.J. Neal. 1996. *Living by the Rules of the Sea*. Durham, NC: Duke University Press.

Campbell, A.C. 1994. *Hamlyn Guide to Seashores and Shallow Seas of Britain and Europe*. London: Hamlyn.

Carter, R.W.G. 1988. *Coastal Environments: An Introduction to the Physical, Ecological and Cultural Systems of Coastlines*. New York: Academic Press.

Carter, R.W.G., and C.D. Woodroffe. 1994. *Coastal Evolution: Late Quaternary Shoreline Morphodynamics*. Cambridge: Cambridge University Press.

Cazes-Duvat, V., and R. Paskoff. 2004. *Les littoraux des Mascareignes entre nature et aménagement*. Paris: L'Harmattan.

Challinor, H., S.M. Wickens, J. Clark, and A. Murphy. 1999. *A Beginner's Guide to Ireland's Seashore*. Sherkin Island, Ireland: Sherkin Island Marine Station.

Collier, M. 2009. *Over the Coasts: An Aerial View of Geology*. New York: Mikaya Press.

Conybeare, C.E.B., and K.A.W. Crook. 1968. *Manual of Sedimentary Structures*. Bulletin 102. Canberra: Australia Bureau of Mineral Resources, Geology and Geophysics.

Correa Arango, I.D., and J.D. Restrepo Angel. 2002. *Geologia y oceanografia del delta del Rio San Juan: Litoral Pacifico Colombiano*. Medellín, Colombia: Fondo Editorial Universidad-EAFIT.

Coulombe, D.A. 1990. *Seaside Naturalist: A Guide to Study at the Seashore*. New York: Fireside.

Davies, J.L. 1964. A Morphogenic Approach to World Shorelines. *Zeitschrift für geomorphologie* 8:27–42.

———. 1980. *Geographical Variation in Coastal Development*. 2nd ed. London: Longman.

Davis, R.A., ed. 1985. *Coastal Sedimentary Environments*. New York: Springer-Verlag.

———. 1994. *The Evolving Coast*. New York: Scientific American Library.

———, ed. 1994. *Geology of Holocene Barrier Island Systems*. New York: Springer-Verlag.

Davis, R.A., and D.M. Fitzgerald. 2004. *Beaches and Coasts*. Oxford: Blackwell Science.

De Andrés, J.R., and F.J. Gracia, eds. 2000. *Geomorphología litoral, procesos activos*. Madrid: Instituto Tecnologico Geominero de España.

Dean, C. 1999. *Against the Tide: The Battle for America's Beaches*. New York: Columbia University Press.

Dean, R.G., and R.A. Dalrymple. 2002. *Coastal Processes with Engineering Applications*. Cambridge: Cambridge University Press.

Defeo, O., A. McLachlan, D.S. Schoeman, T.A. Schlacher, J. Dugan, A. Jones, M. Lastra, and F. Scapini. 2008. Threats to Sandy Beach Ecosystems: A Review. *Estuarine, Coastal and Shelf Science* 81:1–12.

Domm, J.C. 2004. *Formac Pocket Guide to Canada's Atlantic Seashore*. Halifax, NS: Formac.

Emery, K.O. 1960. *The Sea off Southern California: A Modern Habitat of Petroleum*. New York: John Wiley and Sons.

Fish, J.D., and S. Fish. 1996. *A Student's Guide to the Seashore*. Cambridge: Cambridge University Press.

Fitzgerald, D.M., and P.S. Rosen, eds. 1987. *Glaciated Coasts*. San Diego, CA: Academic Press.

Fletcher, C., R. Boyd, W.J. Neal, and V. Tice. 2010. *Living on the Shores of Hawai'i: Natural Hazards, the Environment, and Our Communities*. Honolulu: University of Hawaii Press.

Flor, G. 2004. *Geología marina*. Oviedo, Spain: Servitec.

Fox, W.T. 1983. *At the Sea's Edge: An Introduction to Coastal Oceanography for the Amateur Naturalist*. Englewood Cliffs, NJ: Prentice Hall.

Freire de Andre, C. 1998. Dinâmica, erosão e conservação das zonas de praia [Dynamics, erosion and conservation of beach areas]. Lisbon, Portugal: Commissariat of the Lisbon World Exposition.

Green, R.J. 2010. *Coastal Towns in Transition: Local Perceptions of Landscape Change*. Collingwood, VIC, Australia: CSIRO.

Griggs, G., K. Patsch, and L. Savoy. 2005. *Living with the Changing California Coast*. Berkeley: University of California Press.

Guilcher, A. 1954. *Morphologie littorale et sous-marine*. Paris: Presses Universitaires de Paris.

Haslett, S.K. 2000. *Coastal Systems*. London: Routledge.

Hayes, M.O., ed. 1969. *Coastal Environments—NE Massachusetts and New Hampshire*. Field Trip Guidebook for the Eastern Section of SEPM. Tulsa, OK: SEPM.

Hayes, M.O., and J. Michel. 2008. *A Coast for All Seasons: A Naturalist's Guide to the Coast of South Carolina*. Columbia, SC: Pandion Books.

———. 2010. *A Coast to Explore: Coastal Geology and Ecology of Central California*. Columbia, SC: Pandion Books.

Hayward, P.J., T. Nelson-Smith, and C. Shields. 1996. *Seashore of Britain and Europe*. London: Harper Collins.

Hubal, J., and J. Hubal. 2003. *A Week at the Beach: 100 Life-Changing Things You Can Do by the Seashore*. New York: Marlowe.

Inman, D.L., and C.E. Nordstrom. 1971. On the Tectonic and Morphologic Classification of Coasts. *Journal of Geology* 79:1–21.

Jacobsen, R. 2009. *The Living Shore: Rediscovering a Lost World*. New York: Bloomsbury.

Jensen, J.R., J.N. Halls, and J. Michel. 1998. A Systems Approach to Environmental Sensitivity Index (ESI) Mapping for Oil Spill Contingency Planning and Response. *Photogrammetric Engineering and Remote Sensing* 64:1002–1014.

Johnson, D.W. 1919. *Shore Processes and Shoreline Development*. New York: John Wiley and Sons.

———. 1925. *The New England–Acadian Shoreline*. New York: John Wiley and Sons.

Kaplan, E.H. 1999. *Southeastern and Caribbean Seashores*. New York: Houghton, Mifflin Harcourt.

Kaufman, W., and O.H. Pilkey. 1979. *The Beaches Are Moving: The Drowning of America's Shoreline*. Durham, NC: Duke University Press.

Kelley, J.T., A.R. Kelley, and O.H. Pilkey Sr. 1989. *Living with the Coast of Maine*. Durham, NC: Duke University Press.

Kelley, J.T., O.H. Pilkey, and J.A.G. Cooper, eds. 2009. *America's Most Vulnerable Coastal Communities*. Geological Society of America Special Paper 460. Boulder, CO: Geological Society of America.

King, C.A.M. 1972. *Beaches and Coasts*. 2nd ed. New York: St. Martin's Press.

Klein, A.H.F., C.W. Finkl, L.R. Rörig, G.G. Santana, F.L. Diehl, and L.J. Calliari. 2003. Proceedings of the Brazilian Symposium on Sandy Beaches. Special issue, *Journal of Coastal Research* 35.

Knox, G.A. 2001. *The Ecology of Seashores*. Boca Raton, FL: CRC Press.

Komar, P. 1998. *Beach Processes and Sedimentation*. 2nd ed. Upper Saddle River, NJ: Prentice Hall.

Landrin, Armand. 1879. *Les plages de la France*. Paris: Hachette.

Larson, E. 1999. *Isaac's Storm*. New York: Crown Publishers.

Lencek, L., and G. Bosker. 1999. *The Beach: The History of Paradise on Earth*. New York: Penguin.

———. 2000. *Beach: Stories by the Sand and Sea*. New York: Marlowe.

Martinez, J.O., J.L. González, L.C. Marin, and J.H. Carvajal. 1998. Geomorfologia y aspectos erosivos del litoral Caribe Colombiano. *Publicación Geologica Especial del Ingeominas* 21.

———. 1998. Geomorfologia y aspectos erosivos del litoral Pacifico Colombiano. *Publicación Geologica Especial del Ingeominas* 21.

McLachlan, A., and A.C. Brown. 2006. *The Ecology of Sandy Shores*. 2nd ed. New York: Academic Press.

Michel, F. 1991. *Les côtes de France: Paysages et géologie*. Orléans, France: Editions du BRGM.

Miossec, A. 1998. *Les littoraux entre nature et aménagement*. Paris: SEDES.

Moody, Skye. 2006. *Washed Up*. Seattle, WA: Sasquatch Books.

Neal, W.J., W.C. Blakeney Jr., O.H. Pilkey, and O.H. Pilkey Sr. 1984. *Living with the South Carolina Shore*. Durham, NC: Duke University Press.

Neal, W.J., O.H. Pilkey, and J.T. Kelley. 2007. *Atlantic Coast Beaches*. Missoula, MT: Mountain Press.

Nordstrom, K.F. 2000. *Beaches and Dunes of Developed Coasts*. Cambridge: Cambridge University Press.

———. 2008. *Beach and Dune Restoration*. Cambridge: Cambridge University Press.

Paskoff, R. 1993. *Côtes en danger*. Paris: Masson.

———. 1998. *Les littoraux—Impacts des aménagements sur leur evolution*. Paris: Armand Colin.

Pilkey, O.H. 2003. *A Celebration of the World's Barrier Islands*. New York: Columbia University Press.

Pilkey, O.H., T.M. Rice, and W.J. Neal. 2004. *How to Read a North Carolina Beach*. Chapel Hill: University of North Carolina Press.

Pilkey, O.H., and R.S. Young. 2009. *The Rising Sea*. Washington, DC: Island Press.

Price, A.W. 1947. Equilibrium of Form and Forces in Tidal Basins of the Coast of Texas and Louisiana. *Bulletin of the American Association of Petroleum Geologists* 31:85–111.

Pronzini, E., ed. 1986. *La gestione delle aree costiere*. Rome: Edizione delle Autonomie.

———. 2004. *La forma delle costa. Geomorfologia costiera, impatto antropico e difesa dei litorali*. Bologna, Italy: Zanichelli.

Reineck, H.-E., and I.B. Singh. 1975. *Depositional Sedimentary Environments*. New York: Springer-Verlag.

Ricci Lucchi, F. 1995. *Sedimentographia: A Photographic Atlas of Sedimentary Structures*. 2nd ed. New York: Columbia University Press.

Rothschild, S.R. 2004. *Beachcomber's Guide to Gulf Coast Marine Life*. 3rd ed. Lanham, MD: Taylor Trade Publishing.

Sallenger, A. 2009. *Island in a Storm: A Rising Sea, a Vanishing Coast, and a Nineteenth-Century Disaster That Warns of a Warmer World*. New York: Public Affairs.

Sargent, W. 2008. *Just Seconds from the Ocean*. Lebanon, NH: University Press of New England.

Sayre, A.P. 1997. *Seashore*. New York: Twenty-First Century Books.

Schwartz, M.L., ed. 2005. *Encyclopedia of Coastal Science*. Dordrecht, Netherlands: Springer.

Sept, J.D. 2002. *The Beachcomber's Guide to Seashore Life of California*. Madeira Park, BC: Harbour Publishing.

Shears, R. 2007. "Cappuccino Coast: The Day the Pacific Was Whipped Up into an Ocean of Froth," *Daily Mail*, August 28, 2007.

Shepard, F.P., and H.R. Wanless. 1971. *Our Changing Coastlines*. New York: McGraw-Hill.

Short, A.D. 1988. *Beach Types—Characteristics and Hazards*. Poster, Coastal Studies Unit, University of Sydney, Sydney, NSW, Australia.

———, ed. 1993. Beach and Surf Zone Morphodynamics. Special issue, *Journal of Coastal Research* 15.

———. 1996. *Beaches of the Victorian Coast and Port Phillip Bay*. Sydney, NSW: Australian Beach Safety and Management Project.

———, ed. 1999. *Beach and Shoreface Morphodynamics*. Chichester, UK: John Wiley and Sons.

———. 2000. *Beaches of the Queensland Coast: Cooktown to Coolangatta*. Sydney, NSW: Australian Beach Safety and Management Project.

———. 2001. *Beaches of the Southern Australian Coast and Kangaroo Island*. Sydney, NSW, Australia: Sydney University Press.

———. 2005. *Beaches of the Western Australian Coast: Eucla to Roebuck Bay*. Sydney, NSW, Australia: Sydney University Press.

————. 2006. *Beaches of the Northern Australian Coast*. Sydney, NSW, Australia: Sydney University Press.

————. 2006. *Beaches of the Tasmanian Coast and Islands*. Sydney, NSW, Australia: Sydney University Press.

————. 2007. *Beaches of the New South Wales Coast*. 2nd ed. Sydney, NSW, Australia: Sydney University Press.

Short, A.D., and C.D. Woodroffe. 2009. *The Coast of Australia*. Melbourne, VIC, Australia: Cambridge University Press.

Short, A.D., and L.D. Wright. 1984. Morphodynamics of High Energy Beaches—An Australian Perspective. In *Coastal Geomorphology in Australia*, ed. B.G. Thom, 43–68. London: Academic Press.

Slim, H., P. Trousset, R. Paskoff, A. Oueslati, et al. 2004. *Le littoral de la Tunisie, étude géoarchéologique et historique*. Études d'Antiquités Africaines. Paris: CNRS Éditions.

Snead, R.E. 1982. *Coastal Landforms and Surface Features: A Photographic Atlas and Glossary*. Stroudsburg, PA: Hutchinson and Ross.

Soper, T. 1972. *The Shell Book of Beachcombing*. New York: Taplinger Publishing.

Steers, J.A. 1964. *The Coastline of England and Wales*. 2nd ed. New York: Cambridge University Press.

————. 1969. *The Sea Coast*. 4th ed. London: Collins.

————. 1971. *An Introduction to Coastline Development*. Cambridge, MA: MIT Press.

————. 1973. *The Coastline of Scotland*. Cambridge: Cambridge University Press.

Trenhaile, A.S. 1987. *The Geomorphology of Rock Coasts*. New York: Oxford University Press.

Van Rijn, L.C. 1998. *Principles of Coastal Morphology*. Amsterdam, Netherlands: Aqua Publications.

Vieira, A. 1997. *Praias de Portugal*. Lisbon, Portugal: Editorial Caminho.

Welland, M. 2009. *Sand, the Never Ending Story*. Berkeley: University of California Press.

Williams, A.T., and A. Micalef. 2009. *Beach Management: Principles and Practice*. London: Earthscan.

Witherington, D., and B. Witherington. 2007. *Florida's Living Beaches: A Guide for the Curious Beachcomber*. Sarasota, FL: Pineapple Press.

Woodroffe, C.D. 2002. *Coasts: Form, Process and Evolution*. Cambridge: Cambridge University Press.

Wright, L.D., and A.D. Short. 1984. Morphodynamic Variability of Surf Zones and Beaches: A Synthesis. *Marine Geology* 56:93–118.

Zenkovich, V.P. 1967. *Processes of Coastal Development*. Edinburgh: Oliver and Boyd.

INDEX

COMPOSITION: Westchester Book Group

TEXT: 9.5/14 Scala

DISPLAY: Scala Sans

PREPRESS: Embassy Graphics

PRINTER AND BINDER: CS Graphics